Radar for Technicians

Radar for Technicians

Installation, Maintenance, and Repair

Frederick L. Gould

TAB Books

Division of McGraw-Hill, Inc.

New York   San Francisco   Washington, D.C.   Auckland   Bogotá
Caracas   Lisbon   London   Madrid   Mexico City   Milan
Montreal   New Delhi   San Juan   Singapore
Sydney   Tokyo   Toronto

©1995 by **McGraw-Hill, Inc**.
Published by TAB Books, a division of McGraw-Hill, Inc.

hc   1 2 3 4 5 6 7 8 9 DOC/DOC 9 9 8 7 6 5

Library of Congress Cataloging-in-Publication Data
Gould, Frederick L.
    Radar for technicians : installation, maintenance, and repair / by
Frederick L. Gould.
        p. cm.
    Includes index.
    ISBN 0-07-024062-0
    1. Radar. 2. Radar—Maintenance and repair. I. Title.
TK6575.G65   1995
621.3848—dc20   94-47270
                                                                         CIP

Acquisitions editor: Roland Phelps
Editorial team: Robert E. Ostrander, Executive Editor
                Sally Anne Glover, Book Editor
Production team: Katherine G. Brown, Director
                Wanda S. Ditch, Desktop Operator
                Nancy K. Mickley, Proofreading
                Jodi L. Tyler, Indexer
Design team: Jaclyn J. Boone, Designer   0240620
                Katherine Stefanski, Associate Designer   GEN1

I dedicate this book to my loving wife, Susanne, who has put up with far more than just the long hours involved in writing this manuscript. My very rewarding electronics career that has spanned the last 25 years never would have been possible without her inspiration, support, and advice.

Acknowledgments

A project as involved as a technical book is much more than the efforts of one person. There are many individuals, too numerous to mention, who have helped me in my radar career. Instructors, supervisors, co-workers, operators, and students have all been positive influences in my radar experiences. One technician and supervisor who had the greatest impact on me is George Blaylock, the radar chief at Patuxent River, Maryland. Photographic support was provided by Anita Phillips of Gilfillan, a Unit of ITT Defense and Electronics, and Elizabeth C. Hall, the curator of the Historical Electronics Museum of Baltimore. Finally, the useful technical material that was provided by my good friends Fred Balagna and Hayden Koflowitch was greatly appreciated.

Contents

Introduction

The purpose of this book is to introduce one of the more fascinating areas of electronics—radar. Although the technology has been in application for fewer than 70 years, it has become essential to modern life. Many everyday civilian and military functions would be impossible without the benefit of radar or its associated technology. Radar-based technology has touched our personal lives in the form of weather radar, air traffic control systems, speed guns, microwave ovens, and X-ray equipment, just to name a few. Modern military operations would be impossible without radar's support in threat detection, navigation, and weapons direction. Desert Storm was truly an electronics war as high-tech weapons and search systems revolutionized armed conflict. Used to search for scud missiles and camouflaged armored vehicles, and to guide weapons, radar's abilities were beamed into our homes almost 24 hours a day. Because of radar's impact, the intent of this book is to give you a brief glimpse into a most intriguing aspect of electronics.

The word radar is an acronym for radio detection and ranging. Radar operates on a very simple principle that is familiar: A pulse of energy, when transmitted, strikes a distant object. A very small portion of the energy is reflected back to its point of origin. The received energy is then processed and displayed in such a manner as to give the approximate location of the target. This concept is identical to that of a person shouting into a cave or canyon and hearing an echo. Sound energy travels from the person's mouth, through space, and ultimately strikes a rock surface. A small portion of sound energy is reflected back toward the person. The ear then receives the energy and the echo is heard. Although the concept appears simple, it took many years to develop the necessary technology for it to be useful.

The operation of radar was first demonstrated in 1903 by German scientists using the infant technology to detect the movements of ships at sea. The eminent radio experimenter, Marconi, attempted to interest the British government in radar possibilities in 1922, but he failed. In the 1920s, experiments continued in the United States, Great Britain, Germany, Japan, and other nations. By 1930, U.S. research resulted in the ability to detect high-flying aircraft at short ranges. Many of the early investigations were confined to continuous wave (CW) modulation radar designs. CW systems are able to indicate the presence of an object, but not its distance or accurate position. If accurate range and bearing information is needed, then other types of radar modulation are required.

During the 1930s, most of the developmental work took place in the United States and Great Britain. Hitler's rise to power in Germany gave Great Britain the required incentive to heavily invest in basic radar research. In 1935, a scientist, Robert W. Watt, was approached by the British government about the possibility of fabricating an operational death ray from existing radar technology. His answer was that while a death ray was not feasible at the time, radar could be improved to the point were it could be used to detect aircraft at long ranges. True to his word, with continued governmental support, the state of the art had improved to the point where experimental systems could detect a bomber at ranges of up to 50 miles. Although 50 miles does not sound very far, aircraft speeds at the time were less than 300 miles an hour, resulting in useful warning times for defense. Because it was so promising, research and development continued. The result was the construction of the Chain Home (CH) Radar Stations. The stations were so effective that they were instrumental in the Battle of Britain, which was won by the Royal Air Force. The early warning radar defense system gave the outnumbered British flyers a technological edge that enabled victory. The legacy of the CH stations lives on in the long-range radar warning systems deployed by the NATO and Soviet forces.

Before World War II, radar operation was confined to frequencies ranging from 5 MHz to 20 MHz, which are rather low frequencies by today's standards. At that time, most operational equipment consisted of highly modified standard communications equipment. Research indicated that accurate target detection was dependent on a narrow transmitted beam pattern. At the low frequency of 20 MHz, antenna size is very large. The resulting immense radiating structures are very costly to construct, need a great deal of space, require extensive mechanical maintenance, and are difficult to rotate. Smaller, more manageable antennae required higher frequencies, but the required technology was not yet developed. The essential breakthrough occurred in 1940. Two British scientists, Randell and Boot, developed the cavity magnetron tube. The early device was capable of producing stable, high-powered, high-frequency pulses and was designed to operate between 2 and 3 GHz, capable of producing 1 kW output pulses. Due to shared-technology agreements, American electronics companies were given access to magnetron research, helping to establish an Anglo-American lead in radar research.

Radar might operate on the same principle as an audio echo, but the frequencies involved are much higher. Figure I-1 is a chart that illustrates the different radar bands and frequencies. Even the lowest radar frequency in use, 0.5 GHz, is very high indeed. Because of the frequencies involved, the size and placement of radar components is crucial. At very high frequencies, the interelectrode capacitance that exists between component leads and interconnecting wiring has an effect on circuit operation. Engineers will even use interelectrode capacitance as a circuit component to obtain desired operational characteristics. Antennae, waveguides, and components manufactured for one type of radar often cannot be used on another type, due to different operating frequencies.

Radar band designations	A	B	C	D	E	F	G	H	I	J	K	L	M	
Frequency (GHz)	0.1 0.15 0.2 0.3 0.4 0.5 0.6 0.75 1		1.5	3	5	6	8.0	10	15	20	40	50 60 70 100		
Wavelength (cm)	300 200 150 100 75 60 50 40 30 20 15		10	6	5	3.75	3	2	1.5	1	0.75	0.6 0.5 0.4 0.3		
Communications band designations	VHF		UHF					SHF				EHF		

I-1 *Radar frequency bands.*

Communications users have assigned frequency bands based on the wavelength of the radiated energy. Figure I-1 matches current radar band assignments with corresponding communications bands and frequencies. Communications frequency bands below the radar frequencies are not illustrated. Frequencies from 1 to 300 cycles, or hertz, are considered *extremely low frequency* (ELF). Currently, the only types of equipment operating in this band are the one-way communications systems used to transmit messages to submerged submarines. From 300 hz to about 30 kHz is *very low frequency* (VLF) and is used for communications with satellites and as a backup communications network in the event of nuclear war. The band of frequencies from 30 kHz to 300 kHz is considered to be *low frequency* (LF) and has been used for communications throughout this century.

Medium frequency band (MF) is familiar as the AM broadcast band, and it extends from 300 kHz to 3 MHz. *High frequency* (HF), from 3 MHz to 30 MHz, is used by shortwave broadcasters and intercontinental communications. *Very high frequency* (VHF) extends from 30 MHz to 300 MHz, and 300 MHz to 3 GHz is *ultrahigh frequency* (UHF). Both bands are used by TV, two-way communications, and utility services. Still higher, and now solidly within radar frequencies, is the 3 GHz to 30 GHz range, which is the *superhigh frequency band* (SHF), and 30 GHz to 300 GHz is the *extremely high frequency band* (EHF). These earlier communications designations were not adequate for grouping radar operating frequencies into bands.

Assigning letter designations to radar frequency bands began in WWII, in part to confuse Axis intelligence. Currently, with the exception of the giant Air Force over-the-horizon radars, the lowest operating frequency in use for U.S. radars is the *B band*. That band extends from 0.25 GHz to 0.5 GHz. If you were told that a particular radar, such as the AN/SPY-1 Aegis radar, is an *F band* system, you know that it operates from 3 GHz to 5 GHz.

There are numerous types of radars, and they are classified according to function and type of modulation. The function classification will be discussed first. The most common type of radar is the surface-search radar, with some variations of this type found on ships and boats of all sizes. The equipment provides an operator with bearing and distance information to surface targets, land, and low-flying aircraft. The main use for this type of radar is navigation and observing the movements of other vessels. The Coast Guard operates several shore-based surface-search radar facilities in areas of congested waterborne traffic. Figure I-2 shows a display from this type of radar.

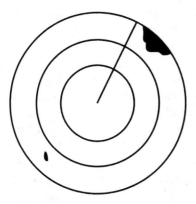

I-2
Surface-search radar display.

The display, or indicator, presents the radar information or video in a meaningful fashion to the operator. Objects such as land masses, aircraft, and ships are presented as a bright shape against a dark background. In Fig. I-2, the large shape in the upper-right side is a typical radar return from a land mass; the small mark in the lower left could be from a ship or low-flying aircraft. Most radar gives very little definition of detected objects. A misconception is that radar can give an almost video picture of objects that requires little interpretation. This type of radar uses a higher frequency, as it must detect echoes from small ships and low-lying land masses. Sailboats and fishing boats are very hard targets to pick up because of a low freeboard and small reflecting surface. To serve as a navigation radar in close proximity to land, the radar system must have a narrow transmitted pulse width that will give it an excellent range resolution and higher range accuracy at short ranges.

There are several characteristics that determine radar performance. *Range resolution* is the ability of a radar to separate two targets on the same bearing, or angular position, from the antenna. It must also have a high pulse repetition rate to give land a high degree of definition. *Pulse repetition rate* is the number of times a transmitter fires in a second. The transmitted beam of energy should have a pattern of energy that is wide in the vertical plane and narrow in the horizontal. The wide vertical dimension helps to compensate for a ship rolling in the ocean. A narrow horizontal beam will give better bearing resolution. Figure I-3 shows a typical search radar beam pattern, depicting the relationship between the horizontal and vertical beam widths. *Bearing resolution* is the ability for a radar to differentiate between two targets on the same bearing or direction, but at different ranges. A radar that provides just bearing and distance information is classified as a *two-dimensional* (2D) system. Most surface search radars operate in the I and J bands.

Air-search radar is used by civilian and military organizations for the detection of aircraft. The Federal Aviation Administration uses both long- and short-range air-search radar for commercial air traffic control. Prior to the widespread use of radar in commercial air traffic control, midair collisions were all too common. An accident occurred in 1959 that convinced society that long-range positive aircraft control was needed. In broad daylight and clear visibility, two airliners collided over the Grand Canyon, a remote site with very little, if any, other flight traffic in the area to cause distractions. The Federal Aviation Administration was expanded so that all commercial aircraft within U.S. borders were under external control at all times. Even with today's more crowded skies, air travel is still very safe, in part due to the contributions of radar.

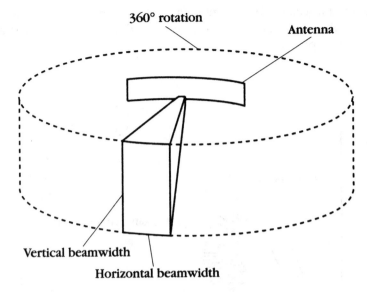

I-3 *Two-dimensional radar beam pattern.*

All branches of the military use air-search radars for both air-traffic control and airborne threat detection. Most air-search radars are 2D and provide range and bearing information of airborne targets. Weather radars also fall into this category. Timely warnings about tornadoes and other violent weather events save many lives and prevent property loss every year. Weather radar sophistication has reached the point where video-processing circuits assign colors to differentiate between intensities of weather phenomena.

Because long range is a requirement, many air-search radars use relatively low transmitter frequencies. To ensure detection of small airborne targets, a long pulse width and high output power are used. The transmitted beam has a wide vertical beam pattern to detect targets from the earth's surface to high altitudes. Many 2D air-search radars operate in the B, C, and L radar bands.

There are specialized air-search radars that provide target range, bearing, and altitude information. These types of radars are called *three-dimensional* (3D). Used for long-range threat detection and obtaining accurate target information for weapons control, 3D radars are very high powered. Because a 3D system uses a lower frequency than a comparable 2D radar, it should have a shorter maximum range. Increasing output power and narrowing the horizontal and vertical beam widths can help the range problem. A 3D radar will transmit numerous narrow beams to obtain altitude information, as pictured in Fig. I-4. As the antenna rotates to provide 360-degree coverage, the beams are stacked from ground level to the highest angle to build a picture.

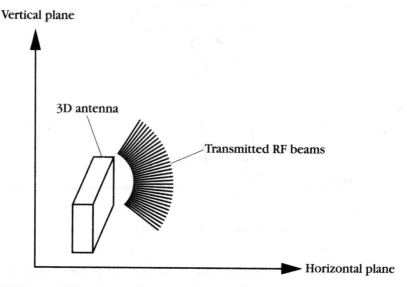

I-4 *Three-dimensional radar beam pattern.*

While most 3D radars use large rotating antennae, a new type of antenna, the *phased array*, is gaining in design popularity. The antenna consists of fixed, immovable radiating faces. Conventional radar uses an antenna that rotates, or is movable in some fashion, which requires a large, heavy, and expensive mechanical pedestal. With the phased array, the radar beam is electronically scanned, which eliminates much of the mechanical maintenance. Another advantage is speed. A mechanical antenna rotational speed is limited by the ability of large gear trains to move the radiating elements. In a phased-array system, a computer moves the beam, which allows the system to more rapidly provide 360-degree coverage. If an object is detected, the radar can immediately direct additional RF beams to paint it, decreasing the time required to investigate and identify it. Many of these radars operate in the E band.

A highly specialized 3D radar, called a *precision approach radar* (PAR), is used to land aircraft during periods of low visibility. This type of radar features extreme accuracy measured in inches. Mounted in a standard military electronics van, a PAR is aligned alongside a runway and provides very accurate range, bearing, and altitude information. A version of this type of radar is mounted on aircraft carriers to provide fully automatic landings. A fully automatic carrier landing calls for high levels of training and confidence between flight crews and operating personnel. Precision approach radar normally operates in the I and J bands.

Weapons-control systems used by the Coalition Forces in Desert Storm illustrate how much radar and other electronic systems have revolutionized warfare. Large numbers of radars were used extensively by both Allied and Iraqi forces. Both shipboard and shore-based missile and gun systems require accurate radar information. The massive, blind firepower of WWII has been replaced with fewer, but far more accurate, shots. For accuracy, the target must be located, and then the weapon/gun system must be directed to the target. Finally, either the missile or bullet stream is guided to the threat. These types of radars have a fast reaction time and extreme levels of accuracy. Weapons direction or fire-control radars produce a very narrow circular beam pattern. Other system characteristics include a high PRF, narrow pulse width, and narrow beam width. They are also called *tracking radars* because the system provides continuous range, bearing, and elevation data to one or more targets. Civilian governmental agencies such as NASA use tracking radars to provide information on space vehicle launches. These radars normally operate in the F, I, and J bands.

Airborne radars are used for navigation, tracking weather fronts, threat detection, and weapons control. These types of equipment have an extreme design problem, as they are more limited by space, weight, and electrical power constraints. The aircraft type determines the characteristics of the radar it will carry. Large airborne search radars such as AWACS and the Navy's Airborne Early Warning aircraft are equipped with long-range air-search radars. Fighters have radars to scan just in front of the aircraft and guide weapons to targets. Commercial airliners have radars to detect weather and other aircraft to prevent accidents.

Another classification describes the transmitted radar signal. The first is *continuous wave modulation*. In this form of modulation, the transmitter outputs a continuous radio frequency oscillation. Due to the nature of CW, transmitter interference with the receiver prevents the detection of stationary targets. The receiver is designed to detect moving targets by sensing a Doppler (or frequency) shift in the received signal. That action is accomplished by mixing or heterodyning a small portion of the transmitted signal with the received echoes. The resulting difference frequency indicates the velocity of the moving target. CW radar is usually found to be low-powered, inexpensive, and short-range equipment. CW systems are predominantly used for radar speed guns, speedometers, and rate-of-climb equipment in aircraft. One variation of the CW radar can give range data. Currently it has limited use as aircraft altimeters.

Pulse-modulation radar can provide the operator with range, bearing, and, in some equipment, altitude information. The system operates by transmitting a series of pulses that are very short in duration, whereas the time in between is much greater. Target range is determined by measuring the elapsed time between the transmitted pulse and the received echo. The bearing or direction of the target is determined by the angular position of the antenna in relation to a reference point. As is the case with many types of electronic equipment, system design is a series of trade-offs. The time between transmitted pulses should be short to provide for accurate positioning of the target. A long time between transmitted pulses is desirable for detecting long-range targets. Short transmitted pulses are required for range resolution, or the ability to separate two targets in close proximity in the same direction from the radar antenna. A long pulse width is used to improve the detection of a weak target. Short pulse width is desirable to provide target resolution. Radar operation is also determined by its frequency. System characteristics (such as range, susceptibility to interference, and types of targets best received) are all influenced by frequency.

From these short pages it is obvious that a well-designed radar system is a series of well-planned but conflicting compromises. The following sections will continue the study of radar basics. Other areas to be presented include the block diagram of a typical radar, transmitter, antenna, waveguide, receiver, and display theory. Additionally, special applications and characteristics will be introduced.

Caution

When following the troubleshooting hints given in this book, always proceed with caution. As a professional you must know and follow all safety regulations to prevent personal injury and equipment damage.

1

Basic radar concepts and installations

As you have already seen, radar and sound wave reflection share several basic principles. As you delve further into the subject, you will also find similarities with optical theory. A single radar transmission has a very limited field of view, possibly as small as 1 degree out of a 360-degree circular field of view. Radar also requires a reference point to enable the operator to interpret the portions of observed objects. For continuity, all search radar antennas rotate in a clockwise direction. Angular measurements of observed targets are referenced to true north, or with ships and aircraft, the bow of the vessel.

Radar circuitry uses trigonometry to calculate target data, such as altitude, angular position, and angle of descent. This concept is illustrated in Fig. 1-1. The earth's surface is represented by a parallel line referred to as the *horizontal plane*. Angles that result from measuring target altitude are the *vertical plane*. The line from the radar antenna directly to the target is called *line of sight* (LOS). The angle formed by the LOS and the horizontal plane is the *elevation angle*. The angular position of a target can be either referenced to true north or the bow of the ship, which is called *relative bearing*. Radar actually uses the coordinates of bearing, range, and elevation to fully describe target location as referenced to antenna position.

From its inception, accurate range measurement has been the most noted radar characteristic. That is possible because the velocity of radiated energy is nearly constant under all atmospheric conditions. Radio frequency (RF) energy travels through free space in a straight line (LOS) at the speed of light, 186,000 miles per second. Al-

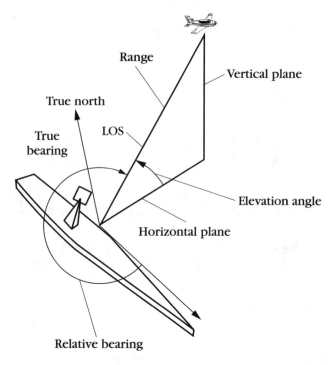

1-1 *Antenna positional information.*

though the velocity of light is in statute miles, radar calculations use the nautical mile, which is slightly longer, 6080 feet versus 5280 feet.

Because RF travels at a constant velocity, distance to a target can be computed by timing how long it takes for energy to leave the antenna, strike a distance object, reflect back, and be received by the antenna. Through experimentation it was discovered that if an object is one nautical mile from a radar, it takes 12.36 microseconds for the RF to complete the circuit. If two-way travel takes 12.36 microseconds, then it would take 6.18 microseconds for the energy to travel one nautical mile. Therefore, a radar mile is said to be 12.36 microseconds in duration. This can also be expressed as the distance a radar pulse travels in one microsecond, which is 320 yards. Because of this constant, all radar signal measurements are expressed in reference to time. A radar system actually measures the distance to an object by measuring the time it takes for the energy to complete the round trip.

One of the more common radars in use today is the pulse modulation type. It functions by transmitting a brief pulse of energy. The transmitter is then disabled to allow the receiver to process any re-

flected energy from an object. The pulse of radiated energy determines several radar system characteristics.

Maximum range is in part determined by the transmitted pulse's
frequency, peak power, repetition frequency, and repetition rate.
Peak power affects range because it determines if enough energy can
be reflected from a target to be processed and displayed on the indicator for operator use. In Fig. 1-2, *peak power* is the maximum amplitude of the transmitted pulse. *Pulse repetition time* (PRT) is the
amount of time, measured in microseconds, between transmitted
pulses. PRT is important because it represents the maximum amount
of time energy that can be received before the next pulse is transmitted. *Pulse repetition frequency* (PRF) is the number of pulses that a
transmitter produces in one second. A lower PRF correlates to fewer
pulses, indicating a longer time period for energy to be received and
processed. PRT and PRF are reciprocals. If one is known, then the
other can be calculated by the formula:

$$PRT = \frac{1}{PRF}$$

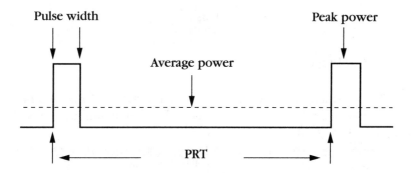

1-2 *Pulse modulation characteristics.*

Carrier frequency is important in terms of range because atmospheric conditions do not affect all frequencies the same. Radars operating on frequencies higher than 8 gigahertz (GHz) are severely
affected by clouds and weather fronts. Taking advantage of this phenomenon, numerous weather radars are in place to provide an accurate weather picture across the entire nation. Most long- and
medium-range radars for both civilian and military use operate in the
3-to-8 GHz range. Frequencies above 3 GHz are attenuated by the atmosphere, which reduces the power that strikes a target and returns
to the system. Atmospheric attenuation affects both the transmitted
pulse and the reflected energy.

Minimum range is another characteristic controlled by the transmitted pulse. In Fig. 1-3, pulse width is the duration of the transmitted pulse at the *half-power points*, the position on a waveform where 50 percent of maximum power is measured. Pulse width, as with other electronic waveforms, is measured in microseconds. Pulse width is minimum range because as long as the transmitter is active and connected to the antenna, the receiver is disabled and unable to receive any reflected energy. Protection circuitry, which allows only the transmitter or receiver to be connected to the antenna, also has a small amount of time that it takes to change position. That is called *recovery time*. Minimum range is the sum of the pulse width and the recovery time.

minimum range = pulse width + recovery time × 320 yards

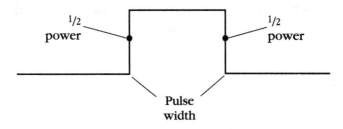

1-3 *Half-power points.*

If a transmitter produces a pulse with a width of 11 microseconds and recovery time is 1.2 microseconds, then minimum range would be calculated as follows:

minimum range = 11 microseconds + 1.2 microseconds × 320 yards
minimum range = 12.2 microseconds × 320 yards
minimum range = 3904 yards

With these parameters, the radar system would be blind to any objects closer than 3904 yards. If the radar is a long-range air search, then this range is acceptable. For a shipboard navigation radar, it would inhibit the radar from providing accurate distance information close to land or other vessels, which could result in collisions and groundings.

A concern of radar operation is ambiguous returns. At the transmission of a pulse, the entire radar system is reset to zero. That action is required to ensure range accuracy to targets. A large target located beyond the maximum range of a radar can still return enough energy to be processed by the receiver. It would arrive at the antenna after the second pulse was transmitted. If a system had a maximum range of 50 miles and a large target located at 60 miles returned enough energy to be picked up, it would appear to be only 10 miles from the radar, an error of 50 miles. These rare targets can be known as *ambiguous returns, second-time-around video*, or *second-sweep echoes*.

Duty cycle is the ratio of transmitter active time compared to the inactive time. A transmitter produces energy for only a brief period of time, followed by much longer dead time when the receiver is active. Duty cycle is calculated by multiplying the pulse width of the transmitted pulse by the PRF of the transmitter. PRF is the number of pulses a transmitter produces in a second. The formula would be:

$$\text{duty cycle} = \text{PW} \times \text{PRF}$$

If the radar had a pulse width of 3 microseconds and a PRF of 800 pulses per second, the duty cycle would be calculated as follows:

$$\text{duty cycle} = .000003 \times 800$$
$$\text{duty cycle} = .0024$$

Power is another system characteristic that must be accurately known. A transmitter has peak power and average power measurements, as illustrated in Fig. 1-2. *Peak power* is the maximum value of radiated power measured in watts produced by the RF transmitter. *Average power* is the value of power produced during one complete cycle of transmitter operation. A cycle consists of one transmitted peak pulse followed by the dead time. Transmitter dead time is when the receiver is active and processing video. The cycle ends when the next pulse is produced. As a result, the average power is much lower than the peak power. Average power is calculated with the following formula:

$$\text{ave. pwr} = \text{peak power} \times \text{PW} \times \text{PRF}$$

Both antenna and target height have a definite effect on a radar's maximum range. As the radar beam is essentially line of sight (LOS), it is unaffected by the curvature of the earth. Rather than follow the surface of the earth, it goes in a straight line into the atmosphere. Figure 1-4 is a pictorial representation of the effect. The radar horizon is the maximum range that a target on the surface of the earth would be visible to the radar. As shown, an island below the horizon is underneath the radar beam and is invisible. However, the aircraft, which is farther away, is visible, as it is at a much higher altitude. The formula to calculate the distance to the radar horizon is:

$$\text{radar horizon} = 1.25 \sqrt{\text{antenna height}}$$

Rotational speed of the antenna must also be considered. The faster an antenna rotates, the shorter maximum range it will have. That is because with a higher speed, less time is spent in each angu-

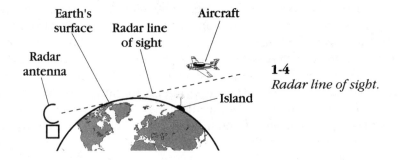

1-4
Radar line of sight.

lar position. If an antenna rotates three times per minute (RPM), it will strike a target twice as many times if it rotated at six RPM instead. A good point to remember is that the more pulses a transmitter radiates in an area, the greater the chance of receiving a large enough return from a target to be displayed, resulting in the ability to detect smaller targets at a greater range.

Resolution is the ability of a radar to separate two or more targets in close proximity. *Range resolution* is when both targets are on the same bearing, but different ranges. Range resolution is determined by factors such as transmitted pulse width, target physical size, and receiver efficiency. A rule of thumb is that a radar system should be able to discern targets separated by one-half pulse width of the transmitter's radiated output pulse. The formula to calculate minimum range resolution is as follows:

range resolution = PW × 164 yards

If a radar has a pulse width of 3 microseconds, the resolution is easily computed. This means that two targets on the same bearing would have to be at least 492 yards apart before the radar would observe them as two separate objects.

Bearing resolution is the ability of a system to distinguish between two targets on the same range, but different bearings. Bearing resolution is directly attributable to the antenna beam pattern. Unlike communications and TV antennas that produce up to a 360-degree radiation pattern, a radar antenna is similar to an optical device such as a telescope. Figure 1-5 is a drawing of a representative radar antenna beam pattern. As shown, the radiated energy forms a long and narrow pattern.

Each time the transmitter radiates a pulse of RF energy, the same pattern is formed. The only difference is if the antenna is rotating, which will cause the beam to rotate. The width of an antenna beam is referenced to the half-power points. The half-power points are one-half the power that is measured at the center of the beam. In Fig.

1-5, the points are marked with dots. Unless a target falls within those points, it will not return enough energy to be displayed as useful information. Beam width can vary from less than 1 degree to over 50 degrees. In terms of bearing resolution, two targets at the same range must be separated by at least one beam width to be displayed separately. Any closer, and the radar will discern them as one object.

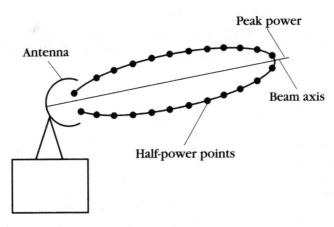

1-5 *Radar beam pattern half-power points.*

Radar installation

In radar installation, location is vital. To properly observe the area of interest, the antenna must have a clear and unobstructed view. The shelter, space, or building that it is installed in must have sufficient space, power, and air conditioning. Systems can be installed on land, on board aircraft, and on ships. Possibly the most challenging installations are on ships and aircraft.

Shipboard installations

The greatest problem encountered in shipboard installation is antenna position. Shipboard radars are typically used for navigation, air traffic control, threat detection, and weapons direction. Installation problems arise because ships are very crowded, and each year more electronic equipment is required. Numerous communications antennae, masts, cranes, other electronic antennae, and rigging all vie for a share of the limited space. Since the beginning of WWII, military ships have become floating forests of antennae.

Figure 1-6 illustrates a typical small warship, a U.S. Navy Parry class frigate. The smallest of all ocean-going combatant warships, it

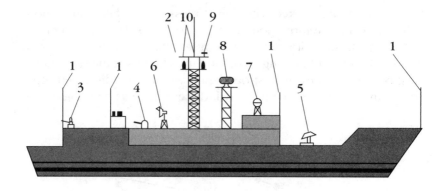

Legend

1. HF communications antennas
2. UHF, VHF communications antennas
3. Phalanx CIWS
4. 76 mm gun
5. Missile launcher
6. STIR radar antenna
7. MK 92 fire control radar antenna
8. AN/SPS-49 long range air search antenna
9. Surface search radar antenna
10. ECM antennas

1-6 *Antenna locations on a U.S. Navy frigate.*

must be equipped with a range of electronic systems. Working within the small space represented by the ship, all antennae, including radar, electronics warfare, and communications antennae, for both reception and transmission must be placed so as to minimize interfering with other systems. Care must also be taken to ensure that radiated energy isn't picked up by rigging and masts. If rigging, masts, and other metal items are not properly grounded, then they can act as receiving antennae and become an electrical safety hazard.

Typically, search radar antennae are mounted on the highest mast or superstructure to give the greatest radar horizon. In Fig. 1-6, located in the center on the mast is the surface-search radar antenna (9). As shown, it is mounted on the highest point on the ship's mast and is provided with a complete 360-degree field of view. Just below, and forward, is mounted the AN/SPS-49 air search antenna (8). The mast will not affect the radiated beam by blocking any energy. That is due to its small size and close proximity to the antenna. As it is an air search, its beam pattern and higher output power still give it a satisfactory range, although it isn't mounted as high on the superstructure. The fire control systems, or radars that direct and control weapons, are mounted lower on the ship's superstructure and are sometimes partially blocked. That is compensated for by maneuvering the ship to provide a clear field of view of any targets of interest, an action known as *unmasking batteries.* The missile control radar (7) is mounted above and behind the missile launcher (5). STIR radar (6) provides targeting information for the gun (4) and the phalanx (3), a close range antimissile system. Sharing the mast with the surface-search radar are the UHF and VHF communications antennae (2) and the electronic countermeasures antennae (10). Electronic countermeasures are used to defeat, blind, or mask enemy radars. Scattered about the ship are several 32-foot whip antennae (1) that are used for HF communications. As shown, antennae are mounted in virtually every conceivable location. This aspect of ship design is an exact science to minimize electronic interference.

The radar equipment itself must be installed as close to the antenna as possible to keep the waveguide runs as short as possible. Waveguides connect the antenna system to the remainder of the radar. Wherever the equipment is installed, there must be sufficient space for equipment to fit and still allow access for maintenance personnel to service and remove large subassemblies. Space for cable runs, waveguides, power panels, and ventilation also must be provided. Poor ventilation has caused many needless problems due to excessive heat and humidity, contributing to shortened component life. Excess humidity, a leading cause of corrosion, damages switches and relay contacts. Equipment cabinets must be mounted so as to not block exits, access to other equipment, and power panels. Surprisingly, that is a fairly common occurrence. In any aspect of installing equipment, preplanning will eliminate many future problems such as having to move equipment cabinets in order to reach other devices or to adjust equipment.

Typical military shipboard installations are spread among several different rooms or spaces. To the casual observer, the only external sign of the installation is the antenna system. On surface ships, such as cruisers, destroyers, amphibious ships, and auxiliaries, the radar transmitter, receiver, video processing equipment, radar switchboard, and antenna switch are located in one space, usually called the radar transmitter room. If multiple radars are installed, such as a surface search and air search, they will often be installed in separate spaces.

Waveguides connect the receiver and transmitter to the externally mounted antenna system. Processed video is routed to the radar displays or indicators that are mounted in several locations. For navigation purposes, the bridge has at least one indicator. That allows the personnel to compare visual sightings with the radar picture. The primary user is the control space, which previously was called the combat information center (CIC) but is now called the combat direction center (CDC). Depending upon ship types, there will be anywhere from 2 to almost 20 indicators. Each one is assigned for a particular use.

Figure 1-7 is a diagram of the CDC on the Parry class frigate depicted in the line drawing. For clarity, some pieces of equipment have been left out of the drawing. This class of ship will have a single indicator or radar display dedicated to surface-search radar (1). The other indicators are essentially computer consoles that display radar data and computer-generated symbols (2). As shown, there are four of them. Each position is dedicated to a different task. All indicators have the capability to display video from both the air-search and surface-search radars. A horizontal radar indicator (5) is provided, as the design allows several people to observe the radar information. To round out the CIC equipment are the consoles on the left. Weapons

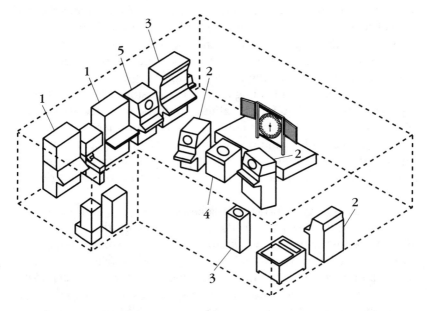

Legend
1. Weapons control console
2. Radar indicator - air search
3. Radar indicator - surface search
4. Radar indicator - horizontal
5. Electronics warfare control console

1-7 *CDC equipment installation.*

control stations (3) for the missiles and gun systems are installed in the same space to allow for the rapid dissemination of information. The final console is the electronics warfare station (4), which is used to confuse or render useless enemy radars. As shown, a modern surface warship is crammed with electronic systems.

Aircraft carriers have an additional space called the carrier air traffic control center (CATCC). Radar video inputs to this space are from a series of air traffic control radars. Long-range air-search information comes from the same radar CDC uses. A 60-nautical-mile radar, the AN/SPN-43, is the marshal radar. It controls arriving and departing aircraft within 60 miles of the ship. The AN/SPN-42, being replaced by the AN/SPN-46, is a fully automatic landing system. It can either be used to provide landing instructions, or if placed in automatic, to actually land the aircraft. The AN/SPN-44 is a high-speed gun. The exact landing speed has to be known to ensure that the aircraft will engage the cables. If its speed is too high, it will snap the cables, too low, and it will fatally hit the stern of the ship, called a *ramp strike*.

Finally is the AN/SPN-41 radar, which serves the same function as the microwave landing system (MLS) on a shore-based air field. MLS must be installed on each individual runway. It transmits two electronic beams, one for azimuth, the other for descent, or glide slope. The aircraft has to be equipped with instrumentation to receive the beams. When both the aircraft and runway are suitably equipped, the pilot can safely land in periods of reduced visibility with no outside assistance.

Shore installations

Shore installations can be almost as challenging. First are the environmental concerns. Although it hasn't been completely proven, there are indications that exposure to low-level radio frequency energy emitted by high-powered radar can cause health problems. In the coming years there will be extensive research in this area. Any radar site must be as isolated as possible. Also, terrain has an impact on radar performance. To provide unobstructed coverage, it shouldn't be placed behind hills or in low spots. A hill blocking the radar radiation pattern would result in a blind spot.

If the equipment is installed in a manufacturer-supplied-and-approved shelter, then there will be far fewer challenges. When equipment is installed in permanent buildings, there must be sufficient planning and care. In addition to the shipboard problems, buildings have a few of their own. Conduits for power cables is one problem. When buildings are constructed, there is pressure to keep costs down. Many times the number and sizes of installed conduit are marginal. Although some money might be saved on construction costs, insufficient conduit space can make future installations prohibitive, as it is often encased in concrete floors. If another system has to be installed, then the floor must be torn up, tubing must be replaced, and then the floor has to be repaired.

Ventilation and cooling are just as important on land-based systems. I have personal experience fighting a corrosion problem caused by humidity. I suffered through intermittent problems, chattering relays, disintegrating metal parts, and degraded equipment performance. In this case, although the ventilation system provided only half the required air flow, a dehumidifier cured the problem.

Another consideration is to keep waveguide length as short as possible. Also, minimize waveguide bends and twists. Any time a waveguide is not a straight run, losses and reflections can be induced that lead to radar performance degradation. Waveguides should also be installed so that personnel can't use them for handholds and supports. It is also helpful to paint "DO NOT STRIKE" on waveguides. Resulting dents and small bends can lead to improper operation.

There are general considerations for all equipment installations. All interconnecting cables must be routed to prevent physical interference with any other systems, doorways, vents, and panels. Always use the properly rated cabling for power. Underrated cable leads to many problems, including major electrical and fire safety hazards. It is essential that all cables are secured with cable trays, stringers, hangers, or conduit. Any interconnecting cables should be routed as to prevent their use as handholds.

An often overlooked point is cable labels. Factory-assembled cables are usually marked or coded so that the technician can tell what they are to connect. With age and use, labels are lost. Also, cables fail and have to be replaced. Ensure that all cables are properly marked, and when making new cables or replacing missing labels, use the identical markings as shown on the schematics. If none are indicated on the prints, then use a generic type. A good example would be the video cable connecting the radar receiver and display. On the receiver end, mark it "VIDEO TO DISPLAY." On the display end, mark it "VIDEO FROM RECEIVER." Wiring diagrams of how the system is actually interconnected are very important. If fabricating your own cables, the solder type are superior to the solderless. The mechanical strength of the solder connector is superior, resulting in fewer future failures.

Finally, and most importantly, the system must have a proper ground. Any equipment without an adequate ground is a safety hazard to personnel. Missing and poor grounds can cause power supplies to float, leading to component damage. Other circuits can fail to function, with odd symptoms that are time-consuming to correct.

Shore-based radar systems are spread out. Airports, whether military or civilian, have a great deal of open land around the runways. Figure 1-8 is a drawing of a typical air facility. The tower (1) is the control center of the facility. Here, controllers will use both visual means and electronic aids to guide aircraft on the ground and in flight. Weather information is provided by the weather radar (2). An electronic aid called a VHF Omni Range Tactical Air Navigation (VORTAC) (3) transmits radial, electronic sign posts that guide aircraft. UHF and VHF receivers (5) and transmitters (4) are required for communications among controllers, aircraft, and vehicles on the runways. The primary radar installation is an airport surveillance radar (6) to observe all air traffic within a 60-mile radius. Military fields have a highly specialized system called a precision approach radar (8). Its function is to allow controllers to guide aircraft to landing during periods of reduced visibility. Currently installed on civilian fields and beginning to make its appearance in the armed forces is the mi-

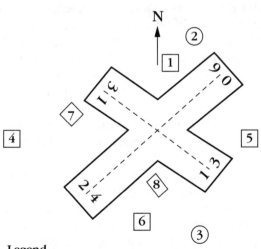

Legend
1. Tower
2. Weather radar
3. Vortac
4. Transmitter site
5. Receiver site
6. ASR
7. MLS
8. PAR

1-8 *Airport radar installations.*

crowave landing system (7). The MLS transmits two guidance beams, one for elevation and the other for bearing data. A readout in the cockpit keeps the pilot on the proper glide slope (descent) and heading (direction).

In addition to large installations such as the ones illustrated, many commercial ships and private boats also have one or two small radars for navigation. Hand-held CW radars are used by law enforcement for speed control. Other applications include using small hand-held units to find buried pipes and conduit. Satellite radars are used for weather, military applications, locating resources, and environmental investigating. The following chapters will discuss circuit theory, operation, and maintenance hints associated with typical radars.

Basic radar block diagram

All radars share many common functions and characteristics. Because of that, a drawing called a basic block diagram can be used to discuss system operation and signal flow. Most technicians use block diagrams when beginning to isolate system failures. Because a system block diagram is very simple and uncluttered, failures can be found faster. Typically, it would be used to isolate the problem to a function. At that point, a much more detailed schematic diagram would be used. This diagram is similar to the functional block diagrams that I have used on U.S. Navy radars such as the AN/SPS-10 shipboard surface-search radar and AN/FPN-63 air traffic control precision approach radar.

Figure 1-9 is a representative radar system block diagram. The operation of any radar system begins in the trigger section, the block in the lower center of the diagram. Triggering has been justifiably called the heart of a radar system. It is also known by other names, such as synchronizer, trigger programmer, or system timing function. The function of the trigger section is to provide system timing to the entire radar system and ancillary equipment. Radars are composed of many separate functions that must operate in precise synchronization. Each function must perform its task exactly when required for the proper time duration. Any deviation will result in a nonoperational and useless piece of equipment.

Output trigger signals are routed to the other radar functions such as the modulator, transmitter, receiver, and displays. Often the system must also provide triggers to ancillary equipment that operates in concert with it. Equipment, such as a video mapper, requires radar triggers. A *mapper* is an electronic device that generates an electronic map to overlay on the radar displays. Operators use it to provide ref-

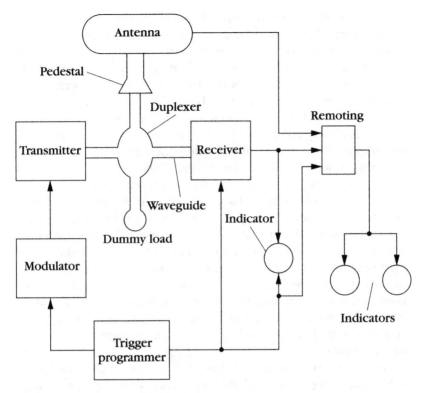

1-9 *Representative radar block diagram.*

erence points to aid in controlling aircraft. Secondary radar, or identification friend or foe (IFF), must also be triggered by a radar. IFF interrogates aircraft and ships with an electronic code. If the aircraft is equipped with the proper equipment, it responds with a coded reply. The reply is then displayed on the radar indicator. The radar video and the IFF replay must be properly placed to ensure accuracy. Civilian authorities use it as an aid in tracking aircraft. In addition to air traffic control, virtually all military forces use IFF for its original purpose—to identify possible hostile aircraft and ships.

Many newer radar systems are either computer controlled or based on computer technology. Digital circuitry requires a signal called a CLOCK for circuit synchronization. The trigger function will also develop this waveform. One area where a signal of this type would be used is in the receiver for the control of video processing functions, such as MTI. MTI functions use digital circuits for short-term, volatile data storage, video processing, and delay lines. A digital delay line is cheaper, smaller, and easier to maintain than the analog devices it replaced.

The modulator receives input from the high-voltage power supply and trigger functions. The crucial function of the modulator is to provide the correct voltage waveform to the transmitter, which allows the microwave device to generate a high-powered RF output pulse. To perform its function, the modulator uses very high operating voltages and currents. Its output must be of the proper amplitude, duration, and timing to ensure accurate system operation. Protection circuitry is also found in this function to prevent faults from damaging components. Faults would include undercurrent, overcurrent, undervoltage, and overvoltage. Any maintenance performed in this area must be done with the utmost caution to prevent personal injury.

The output of the modulator is applied to the RF transmitter. In the transmitter, modulation pulses are converted into high-powered radio frequency (RF) output pulses of energy. The transmitter provides the carrier frequency that is varied by the modulation pulses. The resulting short, high-frequency bursts of energy are routed to the antenna system. The transmitter is another area that requires caution when maintenance is performed. High operating voltages and currents are present any time the system is operational. Arcs are common problems because of humidity, cable separation, and cables breaking down.

The function of an antenna system is to match the impedance of an RF generating device to that of free space. Impedance matching is important because it allows the maximum amount of power to be radiated into space. Any impedance mismatch results in RF energy reflecting back into the transmitter. Reflected energy is not radiated, resulting in less radiated power. Essentially, antennae match the output impedance of the transmitter to that maximum transmission of energy. Any target within the beam of energy radiated by the antenna will reflect a very small portion of it back to the antenna.

A radar's antenna system must serve two distinct functions. It must radiate the high-powered transmitted pulses and also pick up any reflected energy from targets. A good point to remember is that the transmitted pulse has a peak power measured in the thousands or millions of watts, while the received echoes are in the microwatt range. As the same antenna is used for both functions, it must protect the receiver from the high-powered transmitted pulses. If the antenna did allow the full output of the transmitter to be fed into the receiver, damage would result. Because the received energy is so small, any detected energy must be guided only to the receiver. Any reflected energy that is misdirected to the transmitter represents a loss and might degrade overall system operation.

The proper direction of transmitted and received energy is accomplished by a microwave assembly called a *duplexer*. Newer systems use a slightly different design, called a *four-port circulator*. As most radar antennae must be capable of movement, a pedestal containing mechanical gear assemblies is used to provide the motion. The mechanical gear train must be capable of withstanding high winds and still rotate the antenna at the correct speed. Accurate operation requires an antenna that rotates at the designed fixed speed.

The transmitter, receiver, and antenna systems are interconnected by a system of metal tubes called waveguides. A waveguide is a precision pipe that carries high-powered RF signals with minimum loss. Energy is propagated through them as electromagnetic waves. Due to the high frequencies involved, waveguides must be handled and installed to prevent bending and denting. The waveguide must be matched to the radar it is used with, as its dimensions have an effect on the frequency of RF that it will propagate.

Reflected echoes picked up by the antenna are routed to the receiver section. The receiver has three basic functions. It amplifies the low-energy echoes. Amplification is a design challenge because the small echoes are not much larger in amplitude than the naturally occurring receiver noise. The amplified echoes are then detected to extract target information, such as range, bearing, and altitude. The resulting detected video is then processed for further use in the system. Processing could include a function called *moving target indicator* (MTI). MTI allows the radar to display moving targets while eliminating stationary ones. Special circuits are used to ensure that both large, close targets and distant, small ones are amplified within specified limits for display. Without these circuits, the larger targets could overpower the smaller ones to such an extent that they might literally disappear. Newer technology systems might include circuits to develop synthetic symbols to represent targets. Symbols allow operators to rapidly differentiate between unknown, friendly, and hostile targets.

Processed video from the receiver is routed to the radar display, and it is not unusual for the displays to be used quite a distance from the transmitter and receiver site. On board ship, the distance might only be several hundred feet, whereas shore-based systems are often separated by several thousand feet. Shipboard, a piece of equipment called a *radar distribution panel* is used to allow several radars to provide triggers and video to multiple displays. Such flexibility is an advantage. Air-search radars are capable of detecting land, and surface-search radars can pick up low-flying aircraft. If one system is inoperative, the other can provide some information. Shore-based systems use a subassembly, called remoting or a line amplifier, to interconnect the radar with the displays. Remoting must provide impedance matching and amplification due to the large distances involved. Ten thousand feet is a long way to send five volts of video over coaxial cable. No matter where the displays are installed, they require system triggers, video, and antenna information for accuracy.

The function of the display is to present the radar video in a meaningful way that can be used by the operator. From system triggers, the display will generate range rings that aid in determining approximate target distance. For accurate range determination, an internally generated movable bug called a range strobe is used. A typical radar system will use several radar displays. That allows several operators to use the identical information for different purposes.

1-10 *PAR equipment installation.* ITT Gilfillan, a unit of ITT Defense and Electronics

A complete radar system can be as small as a hand-held speed gun or as massive as the USAF's Over The Horizon (OTH) air defense radars. Figure 1-10 is the interior of a trailer-mounted precision approach radar used by the armed forces. Although it is relatively small, it is complex, as it is based on integrated circuits. With today's level of technology, many functions can be found in moderately sized systems. Regardless of the physical size, all radars are based on the same technology. Modern technology enables users to link several radars together to present an accurate picture covering a large area. Well-known examples are the weather and air traffic control radars that cover the entire continent. The following chapters will introduce you to the individual radar functions and provide maintenance hints.

2

Introduction to radar transmitters

The performance of a radar installation is controlled by several system characteristics that are determined in the design phase. As shown in Table 2-1, many of the system characteristics pertain to the transmitter section. Maximum range is a factor that is determined by transmitter output power, pulse width, shape, pulse repetition frequency, and carrier frequency. Minimum range is controlled in part by the transmitted pulse width and shape. System *bearing resolution*, the ability to separate two targets on the same bearing, is set by the transmitter's PRF and antenna reflector beam width. A final consideration is system cost, which is driven predominantly by transmitter output power. With declining defense budgets, cost is a consideration that will increase in attention. For example, a basic radar suitable for a fighter with a 20-mile range costs about $200,000. To double the radar range to 40 miles, the price tag rises to almost $450,000. To obtain a radar with a range of 100 miles, which would be valuable in combat, the price rockets to 2 million dollars. So to increase radar range by a factor of five calls for a tenfold increase in system price.

The most logical point to begin any in-depth discussion of a radar system is the transmitter function. Figure 2-1 is an expanded block diagram of the representative radar system transmitter. The block on the left is the system timing function. All triggers, clocks, and timing waveforms required by the system for accurate operation are developed by these circuits. From the timing section, triggers are applied to various parts of the radar system, including the receiver, radar indicators, ancillary equipment, and the modulator, where the high voltage and current waveform are generated. The modulator, in conjunction with the timing circuits, determines the PRF of the radar system. The high-voltage pulse is routed from the modulator to the RF

Table 2-1 The effects of radar system characteristics on performance

System characteristic	System performance
Transmitter power	Maximum range
	Equipment physical size
	Cost
Transmitted pulse width	Range resolution
	Maximum range
	Minimum range
Transmitted pulse shape	Minimum range
	Range resolution
	Range accuracy
Pulse repetition frequency	Maximum range
	bearing resolution
Carrier frequency	Target resolution
	Directivity
	Equipment physical size
	Susceptibility to atmospheric conditions
Beamwidth	Bearing resolution
Receiver sensitivity	Maximum range

generator. Application of the modulator pulse drives the RF generator, producing the high-powered output RF pulses that are radiated by the antenna system.

Trigger generator

The most important function within any radar system is the triggering circuitry. A trigger function can be known by other names, such as synchronizer, trigger programmer, system timing, master oscillator,

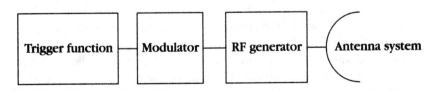

2-1 *Transmitter functional block diagram.*

system oscillator, or keyer. Accurate timing is essential for the proper operation of any radar system. The entire basis for any radar application is the transmission of RF energy pulses at a known time and rate. Elapsed time between pulse transmission and the resulting received energy from objects is then used to determine target parameters such as range, bearing, and altitude. Time comparisons with subsequent echoes from following transmitted pulses are used to ascertain an observed target's velocity and course. Without accurate system timing, a radar's usefulness for detection, navigation, speed measurement, and guidance would be nonexistent. System timing controls the pulse repetition frequency (PRF) of a radar, which in turn determines its maximum range and bearing resolution. *Bearing resolution* is the ability of a system to discern two targets at the same range, but at different bearings. From this, it can be seen that system timing is indeed a crucial function.

Oscillator

All system timing signals and resulting triggers are based on the operation of one type of circuit—some form of an oscillator. An *oscillator* is an electronic circuit that generates a periodic repeating waveform of known amplitude, frequency, and duration. A high degree of stability is mandatory to ensure an output that is constant in frequency and amplitude under all conditions. Oscillations are the result of the controlled periodic storage and release of electrical energy.

One of the earliest types of oscillators, and in many respects, still the best, is the *crystal oscillator.* Its operation is based on a natural phenomenon associated with quartz crystals. It has many advantages, including very stable operation, a high degree of accuracy, inherent ruggedness, and longevity. Crystals have been known to produce accurate oscillations without failure for many years. However, dropping or placing too high a voltage across the crystals renders them useless.

To provide a usable output, the low-level oscillations that a crystal naturally produces must by amplified by external active components. A crystal oscillator can be fabricated with only a few inexpensive external electronic components, such as a transistor, resistors, and capacitors. The resonant frequency of a crystal is determined during the manufacturing process by its physical dimensions. Therefore, it is very difficult to alter its base resonant frequency once it has been manufactured, or "cut." A major disadvantage is that crystals are somewhat temperature-sensitive. If used in a major electronics installation such as a communications or radar system, an oven can be used to maintain the device at a constant temperature to ensure the required degree of stability.

Figure 2-2 illustrates a manufactured crystal's internal construction. Two metallic plates sandwich a very thin slice of quartz crystal. To ensure good mechanical contact, small springs are installed to provide tension. The crystal and metal support parts are hermetically sealed to keep out dust and moisture. Electrical connection to the crystal is accomplished by extending the plates out of the bottom of the case. The plate leads are connected to external support components by either soldering in place or by plugging into a socket. If the equipment has several operating frequencies, a socket mount provides a means to rapidly change crystals. This used to be a common practice in communications equipment.

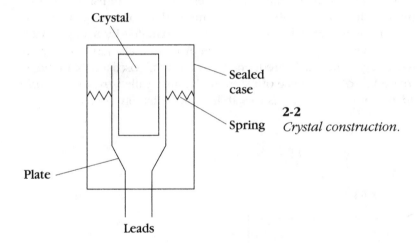

2-2 *Crystal construction.*

Crystal operation is based on the curiosity known as the *piezo-electric effect.* Figure 2-3 depicts the concept. A crystal has two axes, X and Y, that pass through it. The X axis is located through the crystal's corners. Lying perpendicular to the X axis and on the same plane is the Y axis. The action of placing a mechanical stress along the Y axis results in the development of a voltage along the X axis. The process can be observed in reverse by applying a voltage along the X axis, which induces a mechanical stress along the Y axis. The mechanical stress is in the form of naturally occurring vibrations, or oscillations. Through amplification, these oscillations can provide an accurate basis for triggering circuits.

The operation of a typical crystal oscillator can be described using a simple transistor-based crystal Pierce oscillator as drawn in Fig.

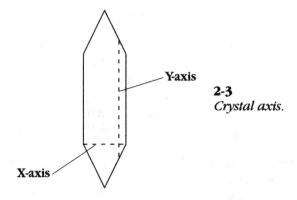

2-3
Crystal axis.

Y-axis

X-axis

2-4. By placing a voltage across the crystal, X-1, it begins to vibrate, or resonate. As the oscillations are very small, the transistor Q1 must amplify them to a usable level. If more than one frequency is required, several crystals can be installed in parallel. By using a switch, the resonant frequency of the oscillator can be changed rapidly and accurately. The basic frequency of a crystal oscillator can be increased through the use of external circuitry, called *frequency multipliers*, but this often leads to stability and reliability problems.

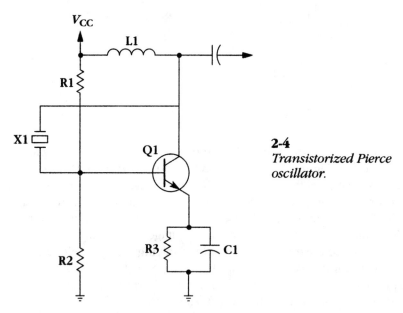

2-4
Transistorized Pierce oscillator.

Another common radar master oscillator is the blocking oscillator. A *blocking oscillator* is a circuit that turns off after producing only one output cycle. It operates for only one cycle because support components reverse bias the input electrode of the active device. Blocking oscillators can be fabricated from vacuum tubes, transistors, or field effect transistors (FET) with a small number of inductors, capacitors, and resistors.

Figure 2-5 illustrates a representative transistorized blocking oscillator circuit. Oscillator operation requires a positive feedback by the coupling between inductors L1 and L2. Coupling is accomplished by having both inductors wound on a common iron core. C1 and R1 form an RC time constant network that determines the resonant frequency of the oscillator circuit. The components labeled R and C are actually the resistive and capacitive components of the inductors and not separate electronic components. A point to remember is that a capacitor will have inductive and resistive components; an inductor will have a capacitive and resistive component; a resistor will have a capacitive and inductive component. The fact that resistance, capacitance, and inductance exist in small degrees in all electronic components due to materials and construction often is not evident until they are used for high-frequency applications. At certain frequencies, the various components might effect circuit operation, either intentionally or unintentionally. Resistor R1 is tied back to the collector voltage, Vcc, to automatically initiate the cyclic circuit operation. Diode CR1 is included to prevent negative transient voltages from placing a forward bias on the collector-base junction and allowing the circuit to function for more than one cycle.

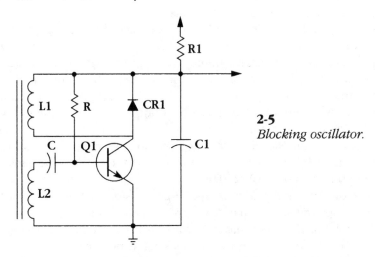

2-5
Blocking oscillator.

With the application of power, the initial flow of collector current through L1 induces a magnetic field. The field in turn cuts L2, inducing current flow through it. L2's current flow shocks the circuit into oscillation. Resonance is determined by L2 and its internal capacitance. On the first positive swing of the oscillation, the base current through Q1 charges the internal capacitance of L2. That places a negative bias on the base input of the transistor, driving it into cutoff and stopping any further oscillations. The internal capacitance then discharges through R, the resistive component of L1. After the charge on C decreases to a low level, the oscillator is once again shocked into oscillation.

Multivibrator

The multivibrator is a very common circuit found in radar equipments for timing purposes. Classified as a relaxation oscillator, the circuit has been in use for timing applications since the dawn of radar technology. The first examples were constructed using vacuum tubes as the active components. As technology advanced, vacuum tubes were replaced by transistors followed by integrated circuits. Circuit operation is based on the charge-discharge time of a capacitive-resistive network interconnecting the two active components.

Figure 2-6 is a schematic of a basic transistor multivibrator circuit. This example is classified as a free-running multivibrator because with the application of power, the circuit begins to function automatically, producing a square wave output of a constant frequency and amplitude. Notice that the two halves of the circuit are a mirror image of one another. Opposite components will have the same value, such as C1 and C2, R2 and R7, and R1 and R6. With the initial application of power, one capacitor will charge faster. That is due to the slight variations that exist in any number of components with the same nominal value. The charging capacitor causes the base current of one of the transistors to rapidly increase. That charging action drives the transistor into saturation, decreasing its output voltage to minimum.

Notice that the transistor collectors are tied back to the base of the opposite transistor. Due to that interconnection, the other transistor is cut off by the action of the saturated transistor, increasing its output to maximum, or Vcc. This condition lasts until the charged capacitor discharges through its discharge resistor back to the base of the opposite transistor. That drives the saturated transistor into cutoff and the cutoff transistor into saturation, reversing the output voltage levels. The cycle then repeats itself as long as power is applied to the circuit. The operation illustrates a basic astable multivibrator characteristic. The circuit has only two possible output conditions: conduction and saturation. During operation, the two outputs will always be

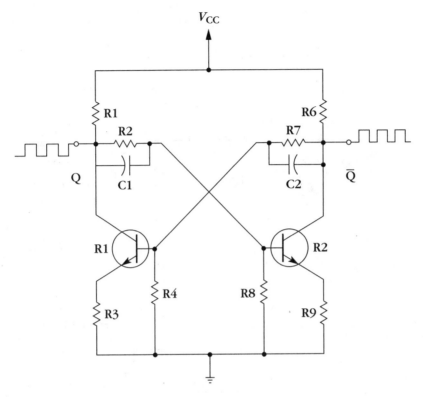

2-6 *Free-running or astable multivibrator.*

opposite; any other combination indicates a failed circuit. When one is high or cut off, the other one has to be low or saturated. These conditions are referred to as states, and the outputs can be called "high" and "low," "on" and "off," or "1" and "0." This type of multivibrator is called an *astable multivibrator* because it constantly changes states, or free runs, as long as power is applied.

Other multivibrator circuits include the *monostable* and *bistable* types, both of which require a trigger for circuit operation. Figure 2-7 is the schematic drawing and timing diagram for the transistorized monostable multivibrator. The major difference between this and the preceding astable circuit is the addition of an input on the lower-left side of the sketch. This circuit requires an external trigger to function. Triggered multivibrators are used to modify the basic system trigger in time for the control of various functions throughout the radar system. When quiescent, or static, the output from the circuit is Q HIGH, Q(NOT) LOW.

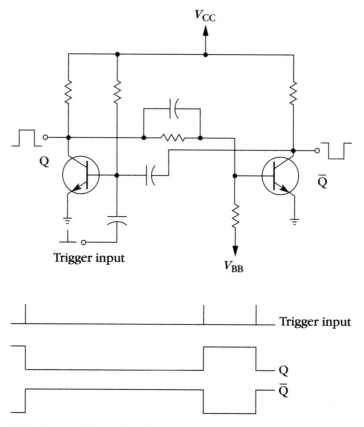

2-7 *Monostable multivibrator.*

That is because the circuit is designed for Q1 to be cut off and Q2 conducting. The application of an input trigger brings Q1 into conduction, driving Q2 into cutoff. Due to the value of the RC components in Q2's base circuit, this condition lasts for a predetermined length of time. At the conclusion of the timing period, the capacitor discharges, returning the circuit to its stable condition. It will remain this way until the application of another external trigger. If a trigger is received during the timing process, the circuit is unaffected. This type of circuit is also known as a *one-shot*.

The bistable multivibrator, illustrated in Fig. 2-8, is a relaxation oscillator that has two stable states. In other words, when power is applied, one output goes high, while the other goes low. It will remain in that condition until the application of an external trigger. At that time it will change states and stay in that stable condition until the application of a second trigger, which will cause it to change states

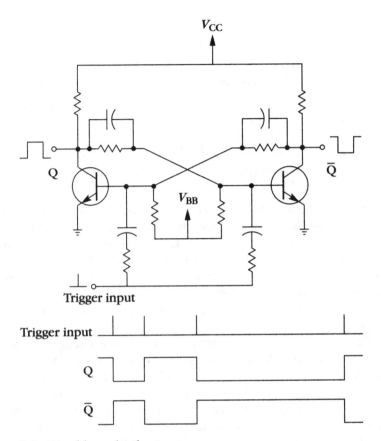

2-8 *Bistable multivibrator.*

again, returning to the original condition. Whatever type of multivibrator is used, they all share several very important characteristics: easy to fabricate, low cost, and stable operation.

Increasingly, modern radar systems are using computers and their basic technology in both radar design and ancillary radar processing circuits. Found in timing, video processing, and control circuits, their application has revolutionized the field. Now it is common to encounter such computer functions as shift registers, random access memory (RAM), and central processing units (CPU). For proper operation, these types of circuits require a synchronization signal called a *clock* to operate. A clock signal must have an exceptionally stable operating frequency due to the tolerances involved.

Figure 2-9 illustrates two typical clock pulses. The waveforms can be *symmetrical,* which is when both positive and negative alternations are equal in time duration. An *asymmetrical clock* is when one

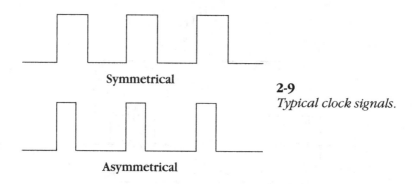

Symmetrical

Asymmetrical

2-9
Typical clock signals.

alternation is greater than the other. The more complex a radar system, the more clocks that are required to synchronize the entire system. In the early 1970s, the AN/TPX-42 IFF system was being installed by all branches of the armed forces and the FAA. The system was so advanced at the time that literally hundreds of clocks and signals were used by the system to ensure accuracy. As you progress in the field, you will find that a clock producing circuit is nothing more than a digital timer or multivibrator.

The vacuum tube and transistor multivibrator circuits are being superseded by digital gates connected as stable, two-output producing circuits, such as the 555 integrated circuit. The reasons for the changes are that as the active devices decreased in size, they also decreased in cost and power consumption. Also, reliability greatly increased, and circuit design and fabrication became easier. The result is that equipment sizes and cost have declined in relative terms while capabilities have dramatically improved.

Figure 2-10 compares the relative sizes of a vacuum tube, transistor, and an integrated circuit (IC). While the tube was replaced on a

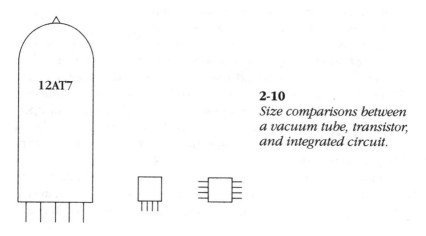

12AT7

2-10
*Size comparisons between
a vacuum tube, transistor,
and integrated circuit.*

one-for-one basis by transistors, that was not the case with ICs. ICs, or chips the size of your thumbnail, contain the equivalent of dozens of transistors, resistors, and capacitors. In addition to the small size, the chips are more efficient in terms of power consumption and heat generation. The digital circuits constructed from ICs have all the advantages of the transistor multivibrator and in addition are smaller, require fewer components, and are more flexible as timers.

Figure 2-11 is the schematic diagram for a monostable multivibrator constructed from integrated circuits. In this example, two digital gates, a capacitor, and resistor form the circuit. Compare that to the transistorized monostable multivibrator depicted in Fig. 2-7. Just on the size basis alone, digital circuitry is far superior to vacuum tubes and transistors. As you study actual radars, you will encounter other digital multivibrator circuits, called *flip-flops*, which are found in many radar timing functions.

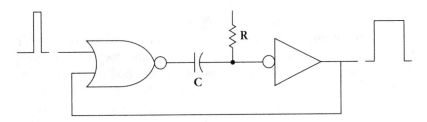

2-11 *Monostable multivibrator.*

The 555 chip, the most common integrated circuit, is an eight-pin dual in-line package (DIP). Figure 2-12 is the drawing of a basic 555 timer circuit. Notice that for operation it requires only four external support components. Capable of operation with power supply voltages that range from 5 to 18 Vdc, it is a very versatile circuit. External components determine the frequency of the timer's output waveform. With just dc voltages, a 555 timer can produce a stable square wave output. By adding an adjustable resistor, the timer is capable of operation over a range of frequencies instead of one fixed frequency.

Circuit operation is straightforward. C2 is a small value component installed to increase circuit stability. R1, R2, and C1 determine the operational frequency of the chip. When power is first applied, C1 begins to charge. As charging begins, the input from C1 is applied to pin 2, the circuit's trigger input. Because C1 is just beginning to charge, the input on pin 2 is low. In this chip design, a low is an enabling input, so on pin 3, the output is active, going high. As C1 con-

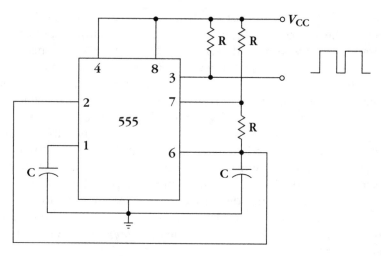

2-12 *555 integrated circuit timer.*

trols the 555, when its charge voltage reaches two-thirds of the power supply, the internal circuitry turns on, disabling the chip, driving pin 3 low. C1 begins to discharge through pin 2 and the 555 timer's internal circuitry. When the charge on C1 reaches one-third that of the dc power supply, the timer is once again enabled, forcing the output on pin 3 to go low. The cycle repeats itself as long as power is applied with a nearly symmetrical, stable square wave as the resulting output.

Digital gates, or flip-flops, have revolutionized multivibrators. Digital multivibrators, or flip-flops, have been engineered using basic gates such as the AND, OR, NAND, and NOR gates. Flip-flops have become very common in modern radars and other electronic equipment because they provide an inexpensive way to obtain accurate timing and storage circuits. Early radars only needed to provide a pretrigger to reset circuits and a master trigger to start system operation every PRT. Modern systems are more complex and, as a result, need a host of triggers and gates to control system operations, which are designed and fabricated from digital circuits.

Figures 2-13A and 2-13B show the schematic symbols for four of the more common digital timing circuits that you will encounter. The first digital multivibrator is the *RS flip-flop*. Essentially a bistable multivibrator, these circuits have only two possible outputs, and they change, or flip, between them. That is why the term flip-flops has come to represent digital multivibrators. The outputs, as with the multivibrators, can be known as high and low, on and off, or 1 and 0.

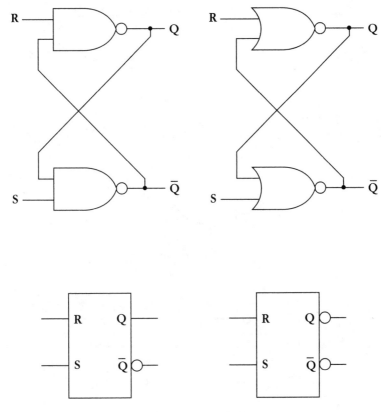

2-13A *Digital RS latch schematic symbols.*

The RS flip-flop is constructed from two NAND gates. The inputs are S and R, and the outputs are Q and Q(not). This logic circuit would be used to control certain functions within the system or in fault detection circuitry. Another name for this arrangement is a *latch*. Anytime the two inputs are opposite, the output changes states.

The next flip-flop is the D-type. This one is very common in the timing function of modern radar systems. A timing signal, a square wave called a clock, is the controlling signal. The outputs can change only on the leading edge, or positive transition of the clock input, as illustrated by Figs. 2-13A and 2-13B.

The final digital flip-flop is the J-K. Notice that the digital circuit has two inputs, J and K, and two outputs, Q and Q(NOT). For the outputs to change states, the J and K inputs must be opposite. Circuit

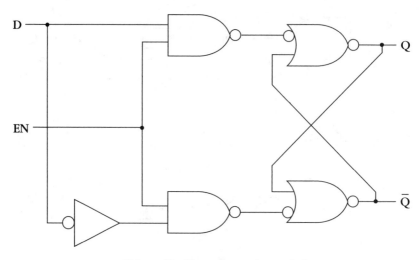

D-type flip-flop schematic symbol

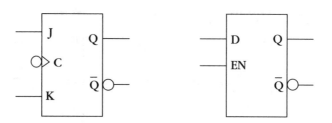

JK flip-flop schematic symbol

2-13B *JK flip-flop and D-type flip-flop schematic symbols.*

operation begins with the application of a clock or timing signal on the CLK input of the JK flip-flop. On the down clock, or negative transition of the clock signal, the flip-flop is enabled, changing the waveforms present on the J and K output pins. For example, assume that statically the J output is high and the K is low. On the negative transition of a clock signal, the ouputs flip, or change states. J goes low and K goes high. On the application of a negative transition on the clock input, the Q output will follow the J input; if J is low, Q goes low. If both inputs are low, then the output remains unchanged with the application of a clock. However, if both are high, then on every negative transition on the clock input, the outputs will change states.

Figure 2-14 is the block diagram for a timing system based on an oscillator circuit. The sine wave produced by the oscillator must be

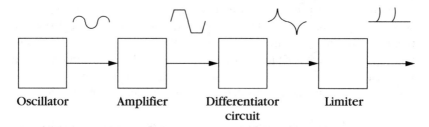

2-14 *Radar system triggers developed from an oscillator.*

amplified to a usable level. The greatly amplified signal is then applied to a differentiator circuit for conversion into a trigger pulse. The positive and negative oscillations are changed into positive and negative trigger pulses. A limiter is then used to eliminate the negative trigger, leaving only the positive one.

The trigger function illustrated in Fig. 2-15 is less complex. A blocking oscillator is used to develop the basic trigger waveform. Because of possible ringing after the main pulse, a limiter is used to pass only the desired positive trigger pulse. The multivibrator-based trigger function is illustrated in Fig. 2-16. The free-running multivibrator is the timing element. A differentiator is used to convert the square

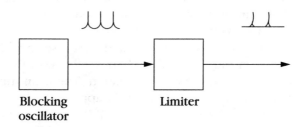

2-15 *Radar system triggers developed from a blocking oscillator.*

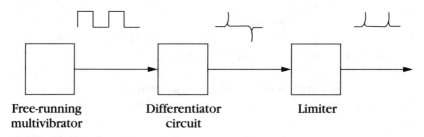

2-16 *Radar system triggers developed from a multivibrator.*

wave into trigger pulses, and the leading edge of the square wave becomes a positive trigger. The negative trigger was formed from the negative square wave transition and is eliminated by the limiter circuit. Of the three possibilities, the multivibrator-based timing function is the easiest to fabricate. Constructed from integrated circuits, it would take up very little space and use a minimum amount of power.

Radar system timing function

To put all of this together, Fig. 2-17 is a simple radar trigger function. The master oscillator is either a free-running multivibrator or some form of an oscillator, such as a crystal-controlled or blocking oscillator. The resulting trigger is the system pretrigger. All other system triggers and clock signals are synchronized to this one signal. Pretrigger is used to initialize the entire radar system. That is required to reset all circuits to stable, or quiescent, states for accurate system operation. It can be used to turn off processing circuits and high-voltage components, and to ensure that there is a sufficient amount of time for high-voltage pulse-forming circuits to begin charging.

The system master trigger is developed by applying the pretrigger to a one-shot multivibrator. The one-shot is used to provide the proper amount of time difference between the pretrigger, which presets all system circuits, and the master trigger, which initiates all operation. The square wave output from the one-shot is applied to a differentiator to convert it to a narrow pulse with sharp transitions. Triggers should have a short duration and rapid rise and fall times to ensure accuracy in triggering other circuits. Pretrigger develops any other system triggers required for operation. The main trigger starts actual system operation. Main trigger is applied to the modulator, where it fires the pulse-forming network (PFN), which in turn causes the RF generator to oscillate, producing an output RF pulse.

The main trigger is used to develop other triggers that control subassemblies such as video processors, MTI circuits, remoting, and internal display triggers. In this example it is routed to three additional functions. The first is to the staggered PRF generator. Staggered PRF is used to increase the blind speed of an MTI radar system. It is also applied to the clock function, which is the digital circuitry that develops all clocks used by the system. The final function is a buffer amplifier. The purpose of this block is to provide output triggers to ancillary equipment such as a video mapper, tower display, and the very important IFF.

Figure 2-18 is a timing chart of the interrelationship that exists between several common signals found in radar systems. As shown in

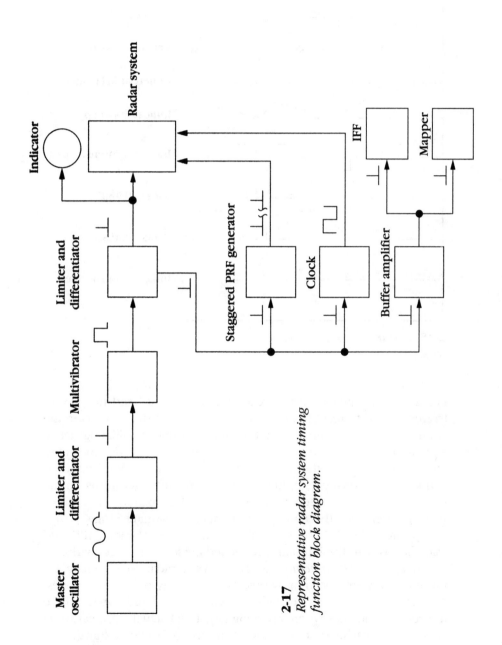

2-17
Representative radar system timing function block diagram.

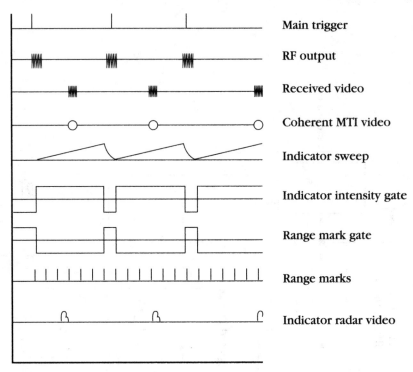

2-18 *Radar system waveforms.*

Fig. 2-18, the signals are interrelated and occur at fixed times. Pre-trigger is used to reset the system to its quiescent state and form the main trigger. The main trigger fires the transmitter, initiating active system action, and starts several waveforms in the radar indicator. Sweep time is when the indicator is capable of displaying returned echoes. The intensity gate, initiated by the main trigger, turns on the CRT, enabling the sweep to be visible to the operator. The range mark gate enables the range mark generator, forming range marks. Composite video is formed by mixing the external radar, IFF, and map videos with the internally generated cursor and range marks.

Moving target indicator (MTI) allows a radar to eliminate re-turned echoes from stationary objects and display only moving ones. A phenomenon associated with MTI circuitry is blind speed. There are certain velocities when a moving target will actually appear to be stationary because of the phase relationship between transmitted pulses and received video. As it has a constant phase relationship, MTI circuits will eliminate the target as if it were a stationary object.

To increase blind speed to the highest velocity possible, a technique called staggered PRF is used. Rather than have one fixed PRF, the system will have several different ones. The system will sequentially step through the various available PRFs in order. The result is an increased blind speed, ensuring that all moving targets are displayed.

Figure 2-19 illustrates a timing diagram for an air-search radar with a staggered PRF. To facilitate understanding the concept, it is compared to a normal PRF. Each trigger pulse is separated by 1000 microseconds from the next pulse. As stated earlier, the basis for any radar system is accurate timing, resulting in a known, constant time interval between trigger pulses. Notice that the staggered PRT cycles through three different pulse repetition times. The first is 750 microseconds, the second is 1000 microseconds, and the third one is 1250 microseconds. The entire radar system is designed to operate under a changing PRT format. In addition to raising blind speed, a side benefit is that a radar with staggered PRF is more jam resistant, a desirable characteristic in military systems.

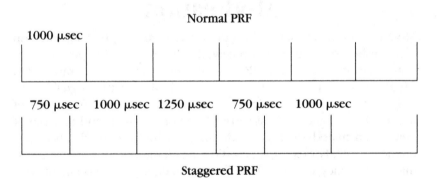

2-19 *Normal PRF and staggered PRF compared.*

Maintenance

Trigger circuits are very easy to maintain and troubleshoot. The most common failure is the loss of one or more trigger waveforms. If pretrigger or main trigger is lost, the most obvious symptom would be a transmitter that doesn't fire. Other symptoms would be all displays blank. First checking the trigger input to the modulator quickly begins to narrow the fault area. If the triggers are good, then either the modulator or transmitter has failed. Missing triggers lead you directly to the trigger function.

If all displays are blank, but the transmitter is firing, then trigger failure would be the most likely suspect. MTI and video processing failures can also be traced to missing triggers in some systems. As many systems provide external triggers to ancillary equipment, their failure could also be a missing synchronization failure.

When checking any circuit used for synchronization, always use internal triggering on the oscilloscope. That way, any triggering problems will not interfere with waveform analysis. Ensure that all triggers and clock signals have the proper amplitude, duration, and frequency. If pulses are to be 800 microseconds apart, and they are only 650 microseconds, then that would cause problems in transmitter output and receiver processing. Circuits such as MTI would be especially vulnerable to timing problems. Clocks can be another source of trouble. Many newer systems use digital circuitry for video delay lines and the control of different receiver functions. As with the triggers, a 1-MHz clock should be just that, not 0.9 MHz. Any error in a clock signal will result in processing problems and many inoperable functions.

Modulation

Modulation directly determines the quantity and type of information that can be obtained from a received echo. Received target information is ascertained by the RF wave characteristics of the transmitted RF. By definition, modulation is the combining of two signals, with the result that the carrier signal voltage is varied by the frequency of the intelligence signal. The carrier frequency is high and capable of being transmitted over great distances, while the modulation frequency is low. For example, consider a talk radio station with a transmitter frequency of 100 MHz. The intelligence is the commentator's voice, which has a frequency of 3000 Hz. The transmitter frequency of 100 MHz is the carrier frequency. The transmitter modulates the 100-MHz carrier with the 3000-Hz intelligence signal. Radar modulation works under the same principle, except with different frequencies. For example, the carrier frequency might be 5 GHz and is modulated by a 600-Hz signal from the trigger function.

A radar system modulates the carrier frequency with a time reference point instead of intelligence. The time of the received signals is compared to the reference point in terms of elapsed time. Time difference between the reference point and the received echo is used to determine target information such as altitude, range, velocity, and elevation. When discussing radar operation, intelligence is the transmitted output pulse, or the main bang produced by the RF generator. To accomplish modulation, a charging network builds up a high-voltage charge. The charge remains stable until the timing section triggers the charging network. That causes the charging network to discharge into the RF section. The sudden high-voltage pulse drives an RF generator such as a magnetron, TWT, or klystron. By design, an RF generator oscillates at a very high frequency, transferring a high-powered pulse into the antenna system for radiation into free space. The output pulse is a square wave composed of the high-frequency oscillations developed by the RF generator. To ensure the time accuracy of received echoes, the rise and fall times should be almost instantaneous.

Radar systems are classified according to the type of modulation that is present in the output RF pulse. The first type of modulation to be presented is *continuous wave* (CW). A CW radar has a continuous transmitter output that is a series of essentially unmodulated RF oscillations. Figure 2-20 compares the various forms of modulation. As shown, a CW output looks exactly like a sine wave. Because it is essentially a sine wave, the signal lacks a reference point, which is mandatory for timing purposes. That means that the radar unit is limited to providing the operator with target-bearing information only. Through processing in the receiver, it is capable of measuring the velocity of a single moving object. It does this by detecting the Doppler shift in received echoes from targets. Velocity measurements are obtained by mixing a sample of the transmitted pulse with the received echoes. Receiver circuitry compares the frequency of the sample transmitted pulse and the received echoes. If the transmitted RF strikes a stationary object, it is reflected back to the radar with no change to the carrier frequency. If the object is moving, the frequency of the reflected energy will be different. The shift in frequency is called the Doppler effect. Therefore, a frequency difference between the transmitted and reflected RF energy indicates the presence and velocity of a moving object.

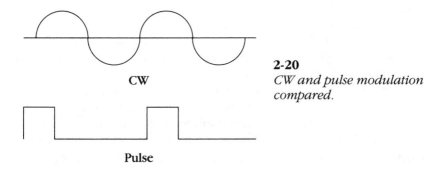

2-20
CW and pulse modulation compared.

CW

Pulse

The best example of the Doppler effect is a train whistle. As the train is approaching, the pitch of the whistle changes from low to high. That is because as the train is coming toward you, the frequency of the whistle increases due to the velocity of the train being added to the velocity of the sound waves. The individual cycles that make up the whistle sound are pushed together due to the motion of the train. That is the same effect as increasing the frequency. When the train passes, and is going away from you, the velocity of the train is subtracted from that of the whistle. The whistle pitch decreased because the train is moving away, stretching the cycles, or decreasing the frequency.

The CW Doppler effect detects the velocity of a moving object by comparing the frequencies of the received and transmitted RF. At radar frequencies, frequency shifts induced by a moving object are very low, in the audio range. Through experimentation, it was discovered that the amount of frequency shift is proportional to the moving object's velocity. With a carrier frequency of 10 GHz, a frequency shift of less than one cycle is enough to obtain an accurate velocity measurement. The difference between the transmitted frequency and the echo is called the difference or *beat frequency*.

Surprisingly, CW radars have numerous and varied uses. Military forces use the technology for proximity fuses to detonate missiles, artillery shells, and bombs. With a CW sensor, the weapon need only be close to the target to explode and cause damage or destruction. Battlefield radars are used to detect the presence of enemy personnel in conditions of reduced visibility. A better-known application is the ever-present police speed guns that we all know so well. Sports groups use them to measure the speed of baseballs, tennis balls, and hockey pucks. Vertical-takeoff aircraft need the technology for rate-of-climb instruments, as downward visibility is extremely limited. CW radars also determine the speed of aircraft landing on board aircraft carriers, which must be known to ensure that an aircraft is landing within velocity criteria.

CW radar has many advantages in providing accurate velocity measurements, and it is inexpensive to manufacture, easy to maintain, and small in physical size. Also, it does not require an expensive high-voltage modulator that is common in other types of radars. As a bonus, it can detect targets at a much shorter range than a modulated radar signal. The minimum range characteristic is much shorter due to the fact that the transmitter and receiver are active at all times. A pulse-modulated radar system must use special tubes or electronic switches to protect the input stages of the receiver from the very high-powered transmitter output pulse, which precludes detecting close targets. Of all the types of modulation available, it requires very little additional circuitry to allow it to distinguish between moving and stationary targets. It does have the major disadvantage of providing bearing information only. The technology is best suited in low-power, inexpensive units. High-powered applications are limited due to the amount of shielding required to protect the receiver from the transmitter's high-powered RF pulses. As with most radar designs, the same antenna is used by both the receiver and transmitter.

FM-CW modulation overcomes the major shortcoming of CW—its inability to measure target range. Range measurement is accomplished by applying a time reference point to the carrier frequency. That is done by varying the carrier frequency as a function of time at a known rate. As with a straight CW system, the frequency of the received signal is compared to a sample of the transmitted pulse. The resulting correlation between transmitted and received frequency gives range and radial velocity information.

The most common use for this type of modulation is as an altimeter in aircraft. The unit is short ranged, with the ground representing a large target. It still has the isolation problem between transmitter and receiver. Also, transmitter noise degrades receiver sensitivity, which can result in a decreased range and ability to detect targets. Other problems are its limited output power, which in turn limits range, and its susceptibility to close-in clutter. *Clutter* is defined as extraneous received echoes from unwanted objects. Ground returns from physical features, wave action on the ocean's surface, vegetation, and buildings can all reflect energy back to the radar, inhibiting a system's ability to detect moving targets. I have found that the unwanted clutter can often return a stronger echo than the desired target.

Pulse modulation is very easy to recognize because it has the characteristic of radiating RF energy in short bursts. The bursts are separated by very long periods of transmitter inactivity, during which the receiver processes any echoes returned to the system. The transmitted pulse duration can range from as short as 0.1 microsecond in duration to more than 50 microseconds, depending on the application the radar is designed for. Modulation pulses developed by the modulator are used to trigger an RF transmitting device that provides high-powered output RF pulse. The pulses of RF energy are radiated to free space by the antenna system. The time difference between transmitted pulse and the resulting received energy provides target range, altitude, course, and velocity information.

The antenna system provides a reference input that determines the angular position of target. The reference point can be true north in the case of shore-based systems. Shipboard and airborne radars can use either true north or relative north. Relative north uses the bow of the ship or aircraft nose as the reference point. Pulse characteristics of this type of modulation have already been discussed.

Figure 2-21 illustrates a complete radar cycle of operation with the waveform characteristics marked. Peak power, average power, PRR, PRT, PRF, and duty cycle are the more common waveform characteristics. *Peak power* is the maximum instantaneous power contained within the transmitted pulse, and in part determines maximum range of the system. *Average power* is the total power that a transmitter produces over one complete PRT, or cycle of operation. As shown, average power is very low in comparison to peak power. *Pulse width* is the duration of the transmitted pulse and determines the minimum range of the radar. *PRT* is the amount of elapsed time measured from the leading edge of the one pulse to the leading edge of the subsequent pulse. The maximum possible range of a radar system is determined by the PRT. PRT, in effect, is the length of time that a radar receiver is active and can process reflected RF energy from a target. *PRF* is the frequency of the transmitted pulses per second. *Duty cycle* is the ratio between the transmitter off time to its on time. All of these characteristics have a direct effect on the information that a radar can extract from a target.

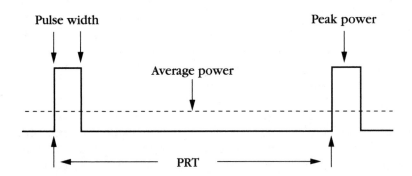

2-21 *Pulse modulation characteristics.*

Doppler radar is a term that is heard every evening during the weather forecasts. A Doppler radar is nothing more than a pulse system that has been modified to employ the Doppler effect to detect moving targets. By choosing different frequencies, a system can be designed to track weather fronts and clouds, or aircraft. A Doppler radar uses coherence. *Coherence* is a phase relationship that exists between the transmitted pulse and the resulting echoes. If the echo is in phase with the reference signal provided by the transmitter, then the object is stationary. A moving object would return an echo that is out of phase with the transmitted pulse.

Pulse radar systems offer obvious advantages over a CW-type system. Any application that requires range, elevation, or altitude information will use pulse modulation. As the transmitter is online for only a brief amount of time, the receiver can be isolated from it. That allows a much higher output power without interference or damage. Greater power translates into better resolution and range characteristics. Disadvantages would be its power consumption, resulting large size, and cost. This technique has many uses, both civilian and military, and is the most common type in use today. Air traffic control, weather radar, threat detection, weapons direction, and observation from space of weather, crops, and natural resources are just a few areas where it is used. Nondestructive examination of crucial parts is a major industry that uses pulse radar technology. A small, portable radar system has been developed that aids in locating pipes and conduit under roads and possible failure points in bridges and roads.

A crucial function performed by the modulator is controlling the RF output pulse developed by the transmitter. It accomplishes this through the application of a modulation pulse of the desired amplitude and duration to the RF generator. To ensure accurate system operation, the modulator and RF output pulses must meet demanding criteria. Pulse shape is similar for both the modulator and RF generator. Figure 2-22 illustrates the proper appearance of pulses and compares it to common failures found in the field. A correct pulse has an almost instantaneous rise time from zero to its maximum value, and it should remain there for the entire time interval that the transmitter is to be producing an RF output. At the end of transmit time, it should decrease to zero almost instantaneously.

The modulator pulse characteristics are required for several important reasons. The leading edge of the transmitted RF pulse is used for range measurements. If the leading edge is indistinct and sloping, the resulting range measurements will be inaccurate and would vary from PRT to PRT. The trailing edge of the transmitted pulse affects minimum range. A steep trailing edge allows objects to be observed closer than if the edge were sloping. The maximum value of the pulse should be constant, as that indicates the most efficient level of transmitter operation. If it varies, that causes a fluctuating transmitter out-

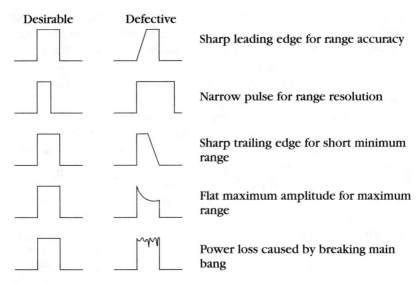

Desirable	Defective	
		Sharp leading edge for range accuracy
		Narrow pulse for range resolution
		Sharp trailing edge for short minimum range
		Flat maximum amplitude for maximum range
		Power loss caused by breaking main bang

2-22 *Transmitted pulse shapes and effects on system operation.*

put power, lowering peak and average power. Decreased power out in turn will affect range resolution and maximum range.

The two most common types of radar modulators are the hard-tube and the line-pulsing. The *hard-tube* is based on the operation of a vacuum tube. Vacuum tube modulators are found in older radar systems and heavy-duty industrial X-ray systems. The vacuum tube functions as a driver. It forms a pulse from the trigger input, which is then amplified and routed to the modulator. Although the arrangement is effective, it suffers from several drawbacks, including lower efficiency, complex circuits, higher operating voltage requirement, and it is more susceptible to line voltage variations. The *line-pulsing* modulator is the more common modulator used in current radar designs. This type uses the same circuit components to store energy and form modulation pulses. It has the advantages of more compact size and less complexity than previous designs.

A line-pulsing modulator consists of a power supply, charging impedance, storage element, and a switching arrangement. The function of the power supply is to provide the correct amounts of current and voltage to the modulator. The time the energy storage element is charged is controlled by the charging impedance. Switching the energy storage element from charge to discharging through the RF oscillator and back to charge is controlled by the switching arrangement. The basis of this function is usually a vacuum tube called a *thyratron*. The most common type of energy storage elements are the pulse-forming network (PFN) and artificial transmission line. The desired output from both is the same, a charge pulse that is rectangular in shape, with sharp rise and fall times, and of the required time duration.

Radar modulator

The radar modulator is considered to be a vital function within any radar system. The primary function of the modulator section is to produce accurately timed pulses of the proper high-voltage amplitude, current, duration, and polarity to allow for accurate system operation. The output pulses generated by the modulator are used as the high-voltage, high-current driving pulse to key the radar transmitter RF generator.

The pulse developed by a modulator must meet a specific criterion. As stated, the pulse must have a specific amplitude, duration, PRT, and shape. The pulse must have steep leading and trailing edges for accuracy and a flat top for maximum power. Peak power of radar is determined by the amplitude of the modulator pulse. Figure 2-23 illustrates desirable modulator pulses. The modulator pulse leading edge must rise from zero to its maximum value almost instantaneously to ensure accurate range measurements. As the transmitted pulse is the minimum range of any radar system, a sharp trailing edge is needed to connect the receiver to the antenna system as quickly as possible. The RF generator is unable to produce maximum power unless the top surface of the pulse is flat. If it is slanted, or even worse, breaking up, output power is greatly reduced. Reduced power decreases system maximum range and the ability to detect small targets, and it induces excessive noise levels into the receiver.

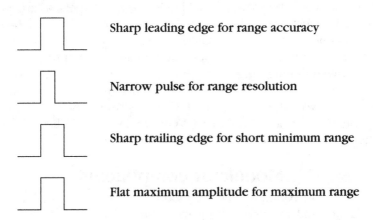

Sharp leading edge for range accuracy

Narrow pulse for range resolution

Sharp trailing edge for short minimum range

Flat maximum amplitude for maximum range

2-23 *Transmitted pulse shapes.*

All radar modulators share several characteristics. For proper operation, they require stable high-voltage and high-current inputs at the proper time interval. The typical circuit consists of a high-voltage switching device, an energy storage component, and protection circuitry. It is common for many radars to be designed with more than one pulse width selectable by the operator to increase system flexibility. For example, short pulse, long pulse, and beacon are frequent pulse widths that are encountered in several radars. Figure 2-24 compares the three different pulse widths.

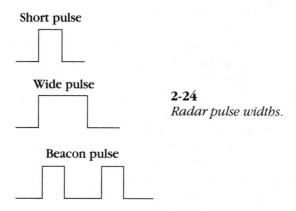

2-24
Radar pulse widths.

Short pulse widths are used for high-resolution applications, such as tracking close range targets, navigation in congested waters, separating multiple targets, or targets obscured by chaff and clutter. Short pulse is considered to be low-powered system operation because the RF generator is enabled for a shorter period of time. Long-pulse-width operation would be used for tracking long-range targets or targets obscured by electronic jamming, as both situations call for high-powered outputs. Beacon is a special-purpose operation that is becoming less common. In this type of operation, a radar system functions as a high-powered, long-range beacon to act as a navigation aid to other ships, aircraft, or personnel on shore. In this instance, a radar system generates two output pulses per PRT. It has been replaced by special equipment such as TACAN and VOR. TACAN is a military beacon, whereas VOR is used by civilian air traffic control organizations.

Modulator components

The two most common types of modulators in use are the hard-tube modulator and the line-pulsing modulator. The hard-tube design is based on a vacuum tube that is used to form the pulse, which is then amplified to a usable level by additional tube. The hard-tube modulator is being replaced, as it has several limitations, including the need for a larger power supply, less efficiency, a more complex design, and high rate of failure. The line-pulsing modulator is more common, as it is simpler in design, more efficient, and uses the same component for energy storage and pulse formation.

Figure 2-25 is an expanded block diagram of a representative radar modulator. Notice that it consists of four major subassemblies: high-voltage power supply, charging impedance, storage element, and switch. The power supply and energy storage element control the system's maximum power. The charging impedance has two functions: to control the charge time of the energy storage element and to prevent short circuiting the power supply during modulator pulse formation. The final component, the switch, fires to discharge the energy storage element into the RF generator.

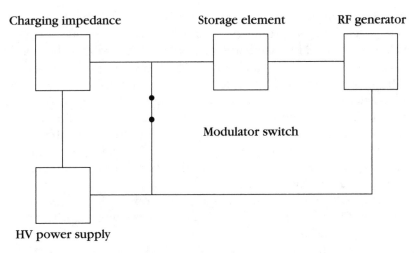

2-25 *Radar modulator block diagram.*

Although the energy storage element is vital to system operation, it is very simple in concept. Depending on the age and sophistication of a system, it can be as simple as a capacitor, or as complex as a massive encased, oil-filled LC network. Capacitive storage elements are found only in radars with a dc power supply, limiting its use to very low-powered applications. The more prevalent devices are either artificial transmission line or pulse-forming networks (PFN). Figures 2-26 and 2-27 illustrate the internal construction of both components.

The function of an artificial transmission line is to store energy between transmitted pulses and, when discharged, form the rectangular modulator pulse. A schematic diagram of an artificial transmission line energy storage element is illustrated in Fig. 2-26. As shown, it consists of a series of LC tanks. The duration of the high-voltage pulse developed by the modulator determines the length of time that the RF generator will produce and radiate an RF output.

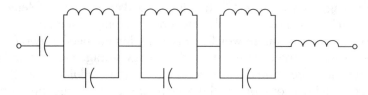

2-26 *Artificial transmission line.*

Modulator pulse duration in the artificial transmission line energy storage element is a result of the number of LC sections and their value. An artificial transmission line is fabricated so that the output end is electrically an open circuit. When voltage is applied, each section, starting from the input side, charges. Discharge begins when the modulator switch closes. The transmission line then discharges through the modulator switch and the primary of the pulse transformer. The discharge action develops a difference in potential. When the potential difference is felt on the output side of the transmission line, its characteristic open is reflected down the entire transmission line, discharging each section in turn.

The pulse ends when the last section on the input side is discharged. Output pulse width is determined by the time it takes the voltage to travel from the input, output, and back again. The time is controlled by the number of sections in the line and values of capacitance and inductance of each section. Because each section of the transmission line feels the full potential of the applied voltage, insulation is vital to prevent possible breakdown and damage.

The pulse-forming network (PFN) is similar to the artificial transmission line, as it is constructed from inductors and capacitors. By examining Fig. 2-27, you will notice that the PFN consists of a series of parallel LC networks. Because of this arrangement, individual capacitors (with the exception of C-1, the input capacitor) do not have to be capable of carrying the full value of applied voltage. That is because the total applied voltage is divided equally among the series of LC networks.

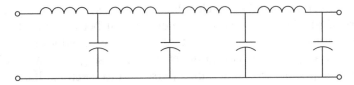

2-27 *Electrical characteristics of a pulse-forming network.*

A PFN is a one-piece, nonrepairable, electronic component. As high voltages and currents are encountered, the unit is immersed in oil to provide insulation, and it is hermetically sealed in a metal case. Figure 2-28 is a photograph of a typical PFN. This one is installed in a Linatron, which is a high-powered commercial X-ray system heavily based on radar technology and components. Notice that external connections are provided to couple energy into and out of the assembly. The connections are the six insulated bolts along the side of the PFN. Markings are provided on the case so that characteristics such as pulse width, impedance, voltage, and current are readily accessible by repair personnel. This is important, as manufacturers often fabricate different radar systems, each requiring unique parts. Also, the end user might have multiple radars to support, so it is imperative that parts are legibly marked.

2-28 *Pulse-forming network.*

Maintenance is limited to cleaning all surfaces and components to remove dust, dirt, and oils. Cleanliness is vital because any foreign matter can form conductive paths to ground. Unwanted ground paths can lead to high-voltage components arcing, causing equipment damage plus exposure of maintenance personnel to electrical hazards. A leaky PFN must be replaced because the oil insulates the capacitors and inductors from the grounded metal case. If you look closely at Fig. 2-28, you'll notice the number of leads that are visible. All of them carry high voltages and currents. Cleanliness is vital, due to the close proximity of conductors.

During system operation, the energy storage device must alternatively charge and discharge. To allow switching between charge and discharge modes, an electronic switch is required. A suitable electronic switch must be able to rapidly switch states and handle high power. First, the electronic switch must close, or go into conduction, in less than a microsecond to allow the PFN to discharge. Secondly, it must open, or cease conduction in less than a microsecond to allow the PFN to begin the charging cycle. While it is rapidly going from conduction to cutoff, it must be subjected to a current flow in the hundreds of amps at a potential of thousands of volts. Finally, it must operate efficiently, consuming a small amount of power.

The requirements are best met by an electron tube called a *thyratron*. Figure 2-29 is a photograph of a typical thyratron vacuum tube. This particular tube is installed in a Varian 3000A, High-Energy Linatron. Just below the tube is a three-position switch to control the value of filament voltage applied to the thyratron. The filaments are selectable to compensate for minor differences between tubes. Ideally, a tube will function with the voltage set to midrange. A thyratron is a gas triode or tetrode that is designed specifically for high-powered switching and control applications. It differs from a conventional vacuum tube in the manner in which the control grid functions. Plate current begins to flow almost instantaneously when grid voltage achieves a particular value. At that time, the grid has no further effect on tube operation. Current continues to flow through the device until plate voltage is either cut off or reverses polarity. Operation should sound very familiar, as it functions the same as a solid-state device, the silicon-controlled rectifier (SCR).

Modulator operation

Basic modulator operation will be covered using Figs. 2-30 and 2-31. Figure 2-30 is the simplified block diagram of a modulator. In this illustration, the modulator switch is open, which allows the storage ele-

2-29
Thyratron deck.

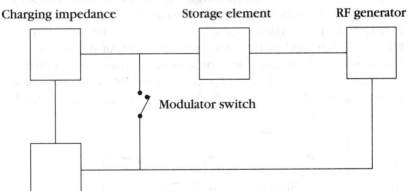

2-30 *Modulator block diagram with electronic switch open and PFN charging.*

ment to begin charging. The charge path is from the storage element, through the charging impedance, the high voltage power supply, and back to the storage element. Figure 2-31 has the modulator switch closed. That condition occurs when the thyratron fires. The discharge

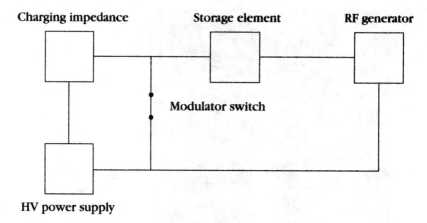

2-31 *Modulator block diagram with electronic switch closed and PFN discharging.*

path is from the energy storage element, through the modulator switch, the RF generator, and back to the storage element.

Actual modulator construction is more complex than four blocks, as can be seen in Fig. 2-32. Function operation begins on the left side of the diagram. The high-voltage power supply provides all voltages required by the assembly. The trigger pulse transformer, the lower-left block, couples the trigger pulse into the modulator and steps it up to a level high enough to trigger the thyratron. The PFN stores the high energy needed to form the modulator output pulse. Two new blocks are the shunt diode and the impedance matching transformer. The shunt diode is vital because often, when a PFN discharges, it will swing negative. That is due to the inherent impedance mismatch be-

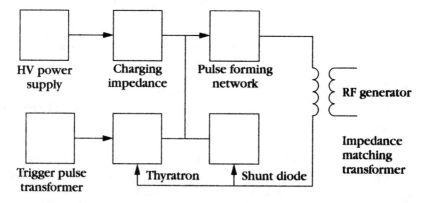

2-32 *Radar modulator block diagram.*

tween the PFN and the RF generator. The final block, the matching transformer, is used to reduce or eliminate impedance mismatches between the modulator and the RF generator.

Because the artificial transmission line and PFN-type modulators function the same, only the PFN will be discussed. Figure 2-33 is the schematic diagram of a representative modulator. After the system is energized, dc high voltage is applied to the modulator, charging the PFN. Current flows from the negative terminal, through the primary of the pulse transformer, the PFN, charging diode, the charging impedance, and back to the positive side of the power supply. The internal capacitance and inductance of the PFN form a resonant charging circuit. With the application of dc high voltage, the PFN attempts to complete a sinusoidal rise to nearly twice the value of the input. After one-quarter cycle, the PFN attempts to discharge as the sinusoidal voltage is decreasing, but as the shunt diode is reverse biased, the charge voltage is maintained. The positive trigger developed by the trigger generator is applied to the primary of the pulse transformer. The transformer secondary develops the high-amplitude pulse required to trigger the thyratron into operation. The application of a positive trigger causes the thyratron to conduct, discharging the PFN. The resulting high-voltage, high-current pulse from the PFN is routed to the RF generator via the pulse transformer.

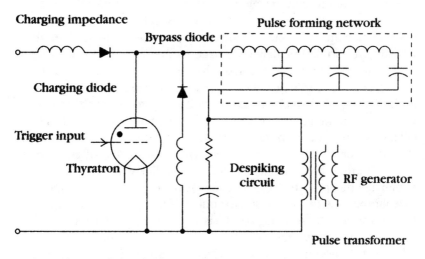

2-33 *Radar modulator schematic diagram.*

The function of a pulse transformer is to step up the output pulse from the PFN and provide impedance matching between the modulator and the RF generator. The output waveform of a PFN is complex, consisting of many high-frequency components. Because of that fact, transformer design is crucial. To ensure that the pulse has a steep leading edge, leakage inductance must be minimized by the use of close coupling between the primary and secondary windings. That is accomplished by winding the primary directly onto the secondary. The secondary is typically *bifilar winding*. A bifilar winding is constructed with two insulated conductors wound next to each other. That results in the same value of voltage being induced in each one. Both windings act as separate secondaries with the same value of in phase voltage induced in each. The advantage of a bifilar winding is that it eliminates the need for high-voltage insulation.

After the PFN is discharged by the thyratron, it attempts to charge negative because of overshoot. That is because there will be an impedance mismatch between the PFN and the RF generator. To prevent the undesirable overshoot, the diode, called the *charge restorer shunt diode*, is forward biased and conducts, totally discharging PFN.

Modulator protection

Although a modulator is designed to withstand high voltages and currents, protection is required to prevent unusual conditions from damaging components. One such condition is *overvoltage*. That happens when the energy storage element charges to a higher-than-normal value. It can cause excessively high pulses to be applied to the magnetron. Such an overvoltage condition can cause the magnetron to internally arc, possibly causing damage to the device. The installation of a spark gap connected across the secondary of the pulse transformer eliminates this problem. Stray capacitance and leakage current in the pulse transformer can cause oscillations to occur after the main pulse fires the RF generator. The resulting negative portion of PFN oscillations can cause spurious outputs from the magnetron. If this should occur, the most pronounced symptom is a loss of close-in targets. A damping diode, connected in parallel with the magnetron, eliminates this problem. The diode is reverse biased during the positive alternations of the modulator pulse. If the pulse goes negative, it is then forward biased and conducts the unwanted pulse to ground.

Despiking is warranted when a spike is present at the leading edge of the modulator pulse. A magnetron has a nonlinear impedance. As a result, under some operating conditions, an impedance mismatch with the output waveguide and antenna might be evident. If a mismatch does exist, the most common symptom is a spike at the leading edge of the pulse. A network to remove, or despike, the waveform must have a resistance that is equal to the impedance of the PFN. The series capacitor must have a very low capacitance so that it will charge rapidly after the PFN output draws full-load current. With the proper selection of network components, any spikes are passed to ground and eliminated.

There are several other types of modulators that are in current use. One is called a hard-tube modulator, illustrated in Fig. 2-34. As shown, this circuit is a vacuum tube that is operated as a class C amplifier. Although it is more complex and expensive, it is one of the more versatile modulators. It can be configured with capacitors and transformers as the coupling elements to the load. It is very flexible in terms of duty cycle and pulse widths. Its main drawback would be its sheer size and cost.

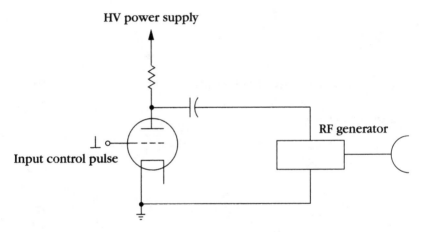

2-34 *Hard-tube modulator block diagram.*

A variation is the floating deck modulator, and it is illustrated in Fig. 2-35. This type of circuit is associated with RF generators such as the traveling wave tube (TWT) and the klystron. Tube 1 and 2 will never be in conduction at the same time. When the RF generator is not producing an output, tube 1 is conducting, and tube 2 is cut off. To bring the RF generator into operation, tube 1 is cut off, and tube 2 is conducting. Gating pulses are coupled to the tubes using either transformers or capacitors.

Solid-state modulators are gaining in popularity. A representative SCR modulator is depicted in Fig. 2-36. The higher cost of solid-state modulators is offset by an increase in reliability. The main limiting factor is a much lower current and voltage-handling capability. The

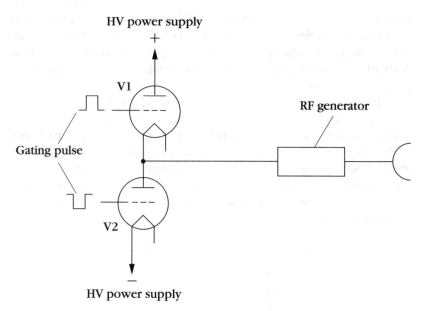

2-35 *Floating deck modulator block diagram.*

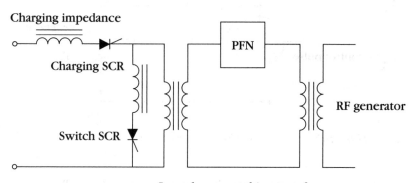

2-36 *Magnetic SCR modulator.*

lower power capabilities can be compensated for by using saturating magnetic cores in series with the SCRs. The time delay inherent with a charging coil limits the current flow through the SCRs until they are completely turned on.

Troubleshooting hints

Any time maintenance is performed in the modulator, extreme caution is required. Figure 2-37 is a photograph of a modulator section. Notice the screen mesh enclosing the chassis. The screen has several

2-37
Modulator cage with shorting probe.

functions. First, it prevents someone from accidentally coming in contact with high voltage. The mesh is held in place by eight screws, so it will take a conscious effort to enter the section. Secondly, it acts as an RF shield, grounding stray emissions to prevent interference with other equipment. To warn personnel of the possible danger, a high danger sign is prominently displayed.

All radar modulators are constructed from components capable of withstanding high voltages and currents. When the equipment is operating, never attempt to place your hands inside of the modulator subassembly. There is never a reason for anyone to do that. Because of the lethal voltages and currents, the presence of your limbs would provide a convenient path to ground, leading to injury or death. If any internal maintenance must be performed, a grounding stick is permanently installed on the cabinet for maintenance personnel. If you look at Fig. 2-37, the grounding stick is the white stick on the right side of the photograph.

If the modulator is the suspected failed subassembly, is it firing? If you have a vacuum tube thyratron, this is easy to determine. A properly operating tube should glow purplish. That is because when triggered into conduction, the tube ionizes. If it isn't glowing, ensure that the modulator is receiving triggers of the proper amplitude and timing. Another good check is to ensure that it has the correct filament voltage. Due to differences in manufacturing, not all tubes will fire with the same filament voltage. Another common failure is the charging diode. Ensure that it isn't shorted or open. If you still have not isolated the failure, check all inputs to the subassembly. That includes triggers, low-voltage power supplies, high-voltage power supplies, and filaments.

Arcing in the modulator is more common than you would imagine. The most common cause is cables and wires too close together. If you suspect this problem, turn out lights to verify where it is occurring. With this type of problem, you must be patient because sometimes the arcs are barely visible. When you do locate the offending point, carefully move the cables. Exercise care because in curing one arc, you might cause another.

RF generators

The function of an RF generator is to produce high-energy output pulses of the required waveshape, frequency, and repetition rate. To accomplish this task, it receives a high-powered input from the modulator and outputs the resulting high-energy, high-frequency pulse to the antenna system for transmission through free space. As always with radar equipment, the technology is constantly evolving, improving RF generator operation and characteristics.

Characteristics of a given radar design are determined by the primary application for the system. These vital and important characteristics include: peak power, average power, radiated pulse length, PRF, stability, distortion, tunability, bandwidth, system cost, useful operational life, efficiency, physical size, weight, and (gaining in importance with every passing year) mean time between failure (MTBF). MTBF is becoming more crucial as it drives support requirements such as spare parts, number of maintenance personnel, and level of maintenance training, all costly line items under constant scrutiny.

RF generating devices

Klystron

The klystron was first developed in the early 1950s. Due to its design, it is capable of a higher peak power than a magnetron, up to 20 megawatts (MW). The high power is possible as the major components comprising the tube are widely separated. The RF tube is an amplifier that can produce a stable output pulse that is controllable in both frequency and phase. In a klystron-based transmitter, it is used as a power amplifier driven by an oscillator. As the oscillator resonant frequency is normally lower than the desired radiated frequency, multipliers are used to increase it to the desired level. The device is capable of producing stable oscillations because the dc and RF sections are separated. Another benefit due to the separation is that the cathode and anode are designed for optimum performance. The inherent stability of this RF device is a highly desirable characteristic in MTI radar systems. There are several drawbacks to the klystron, including expense, large size, mechanical problems, and waste heat produced by normal operation. To compensate for the heat, support equipment (such as a cooling system) is complex, large, and requires additional attention.

Klystron operation is based on electron transit time, considered a detriment in other electronic tubes. *Transit time* is the length of time it takes an electron to travel from the cathode to the plate. The device modulates the velocity of the electrons in transit from the cathode to the plate, forming the stream into bunches.

As illustrated in Fig. 2-38, a klystron can be broken into three sections: gun, RF section, and collector. The gun is composed of heater,

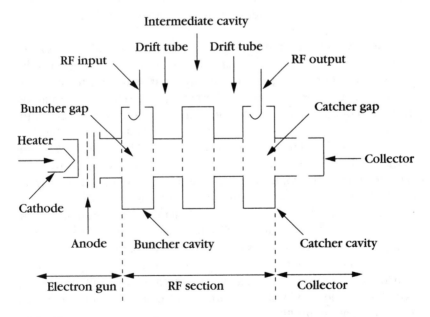

2-38 *Klystron construction.*

control grid, anode, and cathode. The electron beam leaving the cathode is focused by an electrostatic field to ensure narrow beam. The cathode is formed into a concave shape to aid in focusing the beam. A control grid controls the number of electrons leaving the cathode that reach the anode. The grid can be used to turn the tube on and off in pulse applications. By design, the beam is well formed by the time it reaches the anode. Due to the large electron flow and resulting heat, the anode and cathode structures must be large and sturdy to function. The large, high-powered klystrons actually have the electron beam strike the anode with enough velocity to generate X-rays, an undesirable byproduct in most applications. Lead shielding is required for personnel protection. The anode has a hole to allow the electron beam to pass through to the RF section.

The RF section is composed of several resonant cavities, known as resonators. Resonators are connected by metal tubes called drift tubes. The first resonator is the buncher cavity. An RF input is used to modulate the electron stream. The RF input is coupled in by either a waveguide through a slot in the cavity, or by a coaxial cable. To operate, the resonant frequency of the cavity is identical to the RF input. That causes the cavity to oscillate. The oscillations generate an electric field that exists across the buncher gap at the input RF frequency. The positive alternations of the field cause electron velocity to increase. Conversely, the negative alternation causes electron velocity to decrease. Hence, the effect is velocity modulation. The final cavity is the catcher cavity.

The time between the bunches of electrons is equal to the period of the RF input modulation signal. The initial bunch of electrons that strike the catcher cavity sets up oscillations at its natural resonant frequency. That, in turn, causes an alternating E field to be set up across the mouth of the catcher cavity. The bunches of electrons arrive at the proper time to be retarded by the E field alternations. Electron bunches that are retarded, or opposed by the E field, give up energy to the catcher cavity. If you measured the power extracted from the catcher cavity, you would find that it is greater than the RF energy applied to the buncher cavity. Although it seems impossible, it is due to the concentrated bunches of electrons that deliver the energy to the catcher cavity. Additional proof is the fact that the electron beam striking the collector has less energy than the beam leaving the cathode. Not all of the electrons within the klystron are found in the electron bunches. Any unbunched electrons cause a slight power loss from the catcher cavity as they arrive out of phase with the E field across the cavity mouth.

Practical klystrons are usually constructed with more than two cavities, sometimes as many as seven. The additional intermediate cavities are used to improve the bunching action. By increasing the bunching action, gain and tube efficiency are improved. As the cavities are resonant, there are two methods to tune them. Synchronous tuning is when all cavities are tuned to the same frequency, which increases overall device gain but decreases bandwidth. Stagger-tuned is when cavities are tuned to slightly different frequencies to increase bandwidth, but with a decrease in gain.

Klystron operation begins at the heated cathode, which provides electron emission. The electron gun accelerates the electrons and focuses them into a narrow beam. Coils or permanent magnets surround the device, developing a magnetic field that keeps the electron beam focused for the length of the tube. Without the external magnetic force, the electron beam would spread apart due to the individual electrons' natural electrostatic repulsion. The electron beam travels along the drift tube assembly, which connects the resonant cavities. The first cavity provides an input for the RF modulation signal. In klystron action, the electron beam is velocity-modulated by the RF input. In velocity modulation, the instantaneous phase of the input RF will interact with the electron beam, speeding up some electrons and slowing others. The effect is to convert a constant current beam into bunches of electrons. Intermediate cavities have the function of improving the bunching action. The output cavity is placed at the point of optimum electron bunching to obtain the maximum value of RF power.

A klystron is not a linear amplifier. That means that it can be overdriven and saturate, resulting in a decreased power output. Under normal conditions, the RF device provides linear operation up to 70% of maximum output. Klystron saturation is caused by the electron bunches. Maximum klystron power is when the electron bunches are perfectly formed at the output cavity gap. When the RF input is increased above maximum, bunches form too early and begin to spread apart at the output cavity. That is because like charges, negative in this case, repel, so the tightly packed electrons try to move apart.

Klystrons produce a large amount of waste heat. Stray electrons cause heat when striking the drift tubes that join the cavities. Also, the electron beam striking the collector causes heat production. To prevent damage, the tube requires external cooling sources such as heat exchangers or ac.

Total current flow through a klystron is the beam current. Collector current is caused by the electron beam striking the collector. RF or drift tube current is called body current. RF current plus collector current equals beam current. Typically, high-powered klystrons are from 30% to 50% efficient.

Pulsed radars can use high-powered klystrons as RF generators. In this instance, the beam-accelerating voltage is switched on and off. For proper operation, the beam current must also be turned on and off at the same rate. It is very similar to magnetron operation. The RF input can also be pulsed. As with other forms of pulse operation, the beam current must have the period as the RF input. If it isn't pulsed, or has a different rate, then a power loss results due to the constant beam current.

Klystrons have both a fundamental frequency and harmonic frequencies due to the bunches of electrons producing rapid pulses of energy. That is because the tube operates similar to a vacuum tube class C amplifier. The plate current causes oscillations in the output resonant circuits. Harmonics are most noticeable when close to the saturation point. They can be eliminated through the use of harmonic filters. The output can also be distorted due to the nonlinear amplification factor of the tube from 70% to 100% of saturation. A klystron can also generate white noise, which is caused by an electron beam that is not totally homogeneous or by electrons striking the walls of the drift tubes.

As stated previously, the tube is constructed using resonant cavities. The cavities can be tuned through the use of movable walls or paddles that alter cavity geometry. The tubes have proven to be flexible, as frequency agility can be obtained by changing the input RF drive signal. Tuning ranges of up to 20% of center frequency can be obtained without detriment to efficiency.

Traveling wave tube

The *traveling wave tube* (TWT) is a close relative to the klystron. It is a high-gain, low-noise amplifier with a wide bandwidth. Used as a broadband amplifier, it can have a gain of 40 dB or more. Constructed without resonant cavities, the device uses a magnetic field to operate.

Figure 2-39 illustrates a typical TWT. The RF generator uses an electron gun to produce a stream of electrons that is focused into a narrow beam by an axial magnetic field. The magnetic field is produced by a permanent magnet that surrounds the helix portion of the tube. The beam is accelerated as it passes through the helix by a high potential on both the helix and collector. The electron beam continually interacts with the RF electric field along the external circuit. The result is bunching of the electron beam. The area of the tube that contains the electron beam is called the *slow wave structure.*

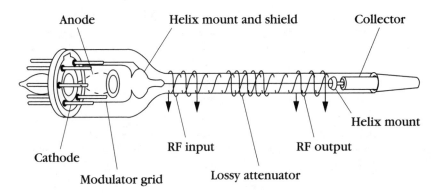

2-39 *Traveling wave tube diagram.*

Amplification occurs when the phase velocity of the propagated wave is in sync with the velocity of the dc electron beam. With current technology, it is difficult to accelerate an electron beam greater than ⅕ the speed of light. Therefore, the forward velocity of the RF field applied to the helix must be reduced to the approximate velocity of the electron beam. Although undesirable, it is possible for electric fields to propagate in either direction along the helix. If it occurs, it can cause unwanted oscillations to occur back along the helix. The spurious oscillations can be prevented or minimized by placing resistive material near the input end of the slow wave structure. The material can be constructed from either a lossy wire attenuator or a graphite coating placed on insulators. A characteristic of lossy sections is that they can completely absorb travelling waves. The electron bunches are unaffected by the lossy sections and are reconstituted by the signal on the helix.

RF has to be coupled into and out of the TWT, as with all RF generating devices. This can be accomplished in one of four methods. The first is waveguide match. A waveguide is terminated with a nonreflecting impedance. One end of the TWT helix is inserted into the waveguide. This method is very efficient because the waveguide has a better Q than TWT. Q is the quality or figure of merit of components such as waveguides, cavities, capacitors, and inductors. A problem is that the TWT bandwidth is greater than that of the waveguide. With this coupling method, as the waveguide has the smaller bandwidth, it is the limiting factor on the RF generator's ability to fully amplify its full potential frequency range.

Cavity match is when one end of the helix is placed in the mouth of a resonant cavity. The helix absorbs any E field energy produced by the cavity. The cavity is designed to resonate over a very wide frequency range. RF energy is fed to the cavity via conventional coaxial cable. Direct coaxial cable coupling has the center conductor of a coaxial cable connected directly to the helix. This method, although very common, has a high VSWR. As the coaxial cable is inserted directly through the glass envelope, heat is generated at the entry point. The final method of coupling is the coupled helix. The coaxial cable center conductor is attached to a small input helix. One end of the major helix is inserted within the small input helix. Electrically, it acts the same as the secondary of a transformer. This method has the advantages of a good VSWR and a broader bandwidth than cavity or waveguide coupling. A drawback is that it cannot handle large amounts of power.

TWT can also be used as RF mixers. Its natural wide bandwidth characteristics can accommodate the frequencies generated by the heterodyning process. A filter on the output of the helix aids in selection of the desired frequency. When configured as a mixer, it has the added benefit of high gain that is associated with a TWT.

A TWT can also be configured as a modulator. In this configuration, the modulation signal is applied to the tube's modulation grid. It has the effect of turning the electron beam on and off. The modulation signal can also control beam density and amplify the modulated output. When utilized as a microwave oscillator, the internal structure of the TWT has to be modified. The oscillator type has a longer helix with no input connection. The operating frequency of the tube is determined by the pitch of the helix.

Magnetron

The magnetron, a breakthrough development in radar, was first developed in 1940. The device has proven to be relatively cheap, rugged, and suitable for portable equipment. As opposed to the klystron and TWT, this device is an oscillator rather than amplifier. Electrically, it is a diode in which the magnetic field that exists between the cathode and plate is perpendicular to an electrical field. Magnetrons are totally self-contained, requiring little or no external support components such as inductors, capacitors, or crystals.

As can be expected, magnetron construction is simple. Figure 2-40 represents a typical magnetron. The tube is mounted inside a permanent magnet with three or four leads protruding from the body of the tube. An external permanent magnet produces a magnetic field of uniform strength that is required for proper operation. Figure 2-41 is the internal construction of a magnetron. The cathode and filaments,

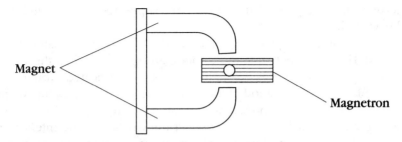

2-40 *Magnetron side view.*

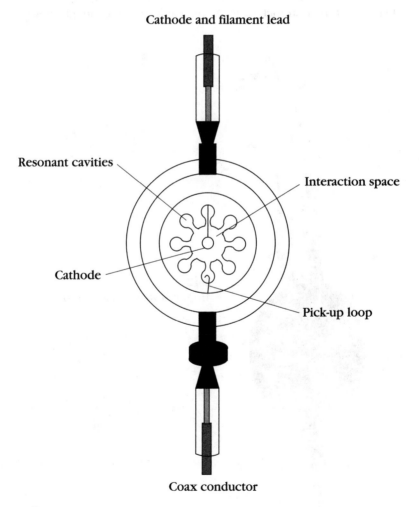

2-41 *Magnetron interior view.*

located at the center of the device, are supported by the rigid and heavy cathode leads.

The anode or plate is constructed from a cylindrical copper block. The space between the cathode and plate is called the *interaction space*. Numerous resonant cavities are fabricated around the circumference of the anode block. The cavities operate as tuned circuits. Slots divide the inner structure into as many separate segments as there are cavities and also connect the cavities to the interaction space. Although not shown in the diagram, alternate segments are strapped together, which connect the cavities in parallel with output. The straps are circular metal bands placed across the top of the anode block at the entrance slots. As the cavities are resonant, they con-

2-42

Magnetron tube. Historical Electronic Museum

trol output frequency of the magnetron. The RF generated by the magnetron is normally coupled out with a probe or loop extending into one of the tuned cavities. It is then routed to the antenna system to be radiated.

Figure 2-42 is a photograph of a small magnetron. The two leads on the right side of the tube are the filament and cathode connections. The connection on the right side is to couple the RF energy out

2-43 *Installed magnetron.*

to the antenna system. The round shape of the tube is dictated by the placement of the resonant cavities around the interaction space. A magnetron appears different when mounted in the equipment. Figure 2-43 is a magnetron installed in a linatron, a high-powered industrial X-ray system. The magnetron is the device mounted across the top of the photograph. If you look on the left side, you will notice the flange bolted to a waveguide. That is how energy is coupled out of this particular tube. Just to the right is mounted the permanent magnet. It is the metal block with a label on it. The label provides information such as its strength, manufacture date, and test date. Across the bottom of the cabinet is the impedance-matching transformer. It was stated earlier that the output impedance of the modulator must be matched with the input impedance of the RF generator to ensure proper operation. You can see that the tube's filaments, located on the far right, are connected to the transformer.

A magnetron tube must be capable of generating high-power pulses. To do so, it must have excellent electron emission characteristics. Much of the output power is derived from electrons emitted when high-velocity electrons fall from the interaction space back to the cathode.

As previously stated, a magnetron is a stable oscillator. Stable means that the tube does not mode shift, double mode, or mode skip. *Mode shifting* is when the magnetron changes from one mode to another during transmit time. The most common causes for this undesirable event is a deteriorated cathode, or an improperly shaped modulator pulse. *Mode skipping* is when a magnetron misfires or fires randomly. The final stability problem is *double moding*. That is when a magnetron will function in two modes during the same modulation pulse.

As mode stability is important to system operation, a method called *strapping* is used to improve tube performance. During the manufacturing process, cavities are interconnected to change their electrical characteristics. Figure 2-44 illustrates the equivalent circuit to an unstrapped magnetron cavity—an LC Tank. The slot connecting the resonant cavity to the interaction space forms an inductor, and the walls of the interconnecting slot form the plates of a capacitor. The result is a high-Q resonant cavity with known frequency characteristics. Electrically, the cavities are connected in series, as shown in Fig. 2-45.

Through strapping, the cavities are connected in a parallel configuration. Figure 2-46 is the resulting equivalent electrical circuit. Strapping is the interconnection of alternate cavities with copper conductors, as illustrated in Fig. 2-47. Notice that there are two separate conductors around the anode block. Through the strapping technique, the space charge that exists in the interaction space rotates around the cathode

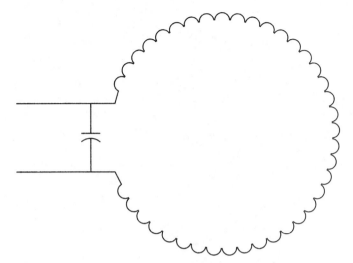

2-44 *Hole and slot cavity magnetron equivalent circuit.*

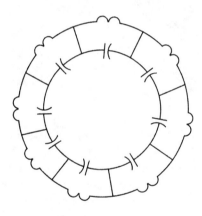

2-45
*Unstrapped cavity
magnetron.
Electrically, the
cavities are in series.*

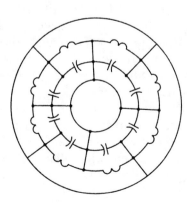

2-46
*Strapped cavity
magnetron.
Electrically, the
cavities are in parallel.*

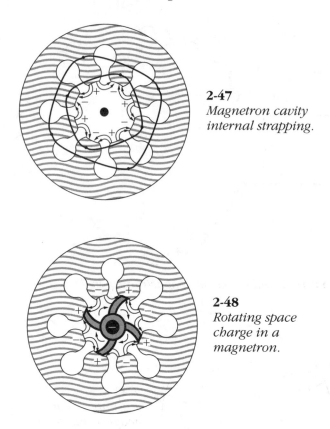

2-47
*Magnetron cavity
internal strapping.*

2-48
*Rotating space
charge in a
magnetron.*

block, with a velocity of two cavities, or anode segments, per cycle of
the applied ac. Figure 2-48 is a drawing depicting the concept. The rotation action aids in sustaining RF oscillations.

Magnetron operation is based on a combination of ac fields and
a dc field. The resonant cavities surrounding the anode block develop the ac fields, as illustrated by the cavity magnetron in Fig. 2-49.
The dc field is the difference in potential between the anode and
cathode. Electrons emitted by the cathode are accelerated by the dc
field and attracted to the anode. In the process of acceleration and at-
traction, the electrons gain energy from the dc field. In the process of
passing through the ac field on route to the anode, the electrons give
up energy to those fields, sustaining oscillations.

As the cavities are strapped, a rotating space charge is developed
by the electrons leaving the cathode and traveling to the anode block.
Some electrons attempt to travel directly. Others, the working electrons, are formed into a rotating space charge wheel that is developed
by the RF oscillations. These electrons can be held for a substantial
amount of time before striking the anode. The space charge wheel ro-

Anode
block

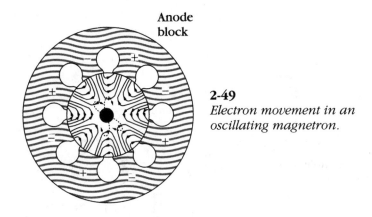

2-49
*Electron movement in an
oscillating magnetron.*

tates around the cathode with a velocity of two cavities per cycle of RF oscillation. Any electrons that are not formed into the wheel are repelled back to the cathode.

During the generation of RF, alternate anode segments separating the cavities are at opposite polarities, which results in the ac field being induced across the mouth of each cavity. As adjoining segments have a phase difference of 180 degrees, this mode of operation is called the pi mode because a 180-degree phase difference is equal to one pi radian. This mode of operation is the most desirable because it produces the maximum possible output power. The pi mode is obtained by strapping alternative cavity segments so that identical polarities are developed. Typical magnetrons will feature six or eight cavities to facilitate strapping. Each strapping ring is at the same potential, but the two rings have opposing potentials. Stray capacitance resulting from the proximity of tube elements adds capacitive loading to the resonant mode. Other modes of operation result in a phase difference between segments connected to a particular strapping ring, which causes current to flow. The straps induce inductance to the magnetron equivalent circuit, resulting in an inductive shunt connected in parallel with the cavity. The effect is to lower inductance and increase the frequency of modes other than the pi mode. The result is that the pi mode is the dominant mode.

There are three common methods of coupling RF out of a magnetron, and they are illustrated in Fig. 2-50. *Loop coupling* is the simplest. A loop projects into one of the resonant cavities. It intercepts the RF energy and couples it to a waveguide. *Segment feed* is when a loop intercepts RF energy between a strap and segment. Coaxial cable then feeds the RF into a waveguide. Aperture or *slot coupling* is when RF energy is coupled directly from a cavity via a slot that directly connects to the waveguide.

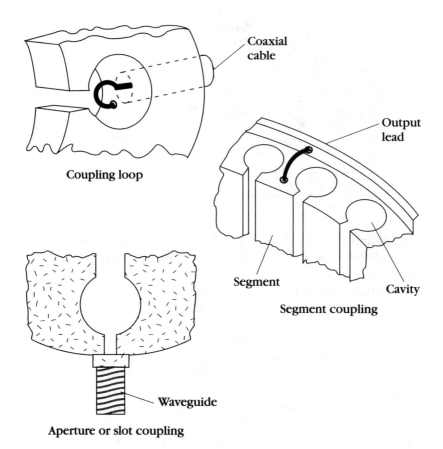

Coaxial
cable

Output
lead

Coupling loop

Segment

Cavity

Segment coupling

Waveguide

Aperture or slot coupling

2-50 *RF energy coupling from magnetrons.*

A variation of the cavity magnetron is the coaxial magnetron. Figure 2-51 compares the internal structure of both types of tubes. All magnetrons exhibit a degree of mode instability because operation is based on resonant cavity oscillations. Device stability can be improved by alternate cavity resonators surrounding the anode block. This design is more expensive due to the manufacturing process, but it features enhanced stability, superior mode control, higher levels of efficiency, and a longer operational life. Based just on the longer life, the coaxial magnetron is cheaper in the long run than the cavity magnetron. During my five years as an instructor in Memphis, Tennessee, we never had to replace a coaxial magnetron in the four AN/FPN63 precision approach radars that were used to train maintenance students. If the radars had been equipped with conventional magnetrons, tube life would possibly have been as short as one year.

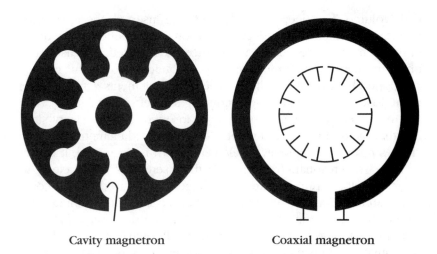

Cavity magnetron   Coaxial magnetron

2-51 *Cavity and coaxial magnetrons compared.*

All magnetrons are capable of operating over a band of frequencies. Because a magnetron is a resonant device, its output frequency can be changed by either altering the capacitance or inductance of the device. The inductance of resonant cavities can be altered by changing the surface-to-volume ratio of the cavity. This type of tuning is called *sprocket* or *crown-of-thorns tuning* and is depicted in Fig. 2-52. Inductive elements are inserted into the cavities to alter

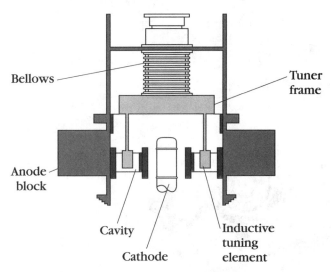

2-52 *Inductive magnetron tuning.*

their volume. The deeper the elements are inserted, the greater the inductance, which in turn decreases resonant frequency. If the depth of the elements is decreased, inductance is decreased, increasing resonant frequency. The tuning elements are attached to a frame and are positioned by a flexible bellows. The tuning elements actually lower the unloaded Q of the cavities, reducing total efficiency.

In capacitive tuning, elements are inserted into the cavity slots, and that method is pictured in Fig. 2-53. Increased element depth decreases resonant frequency. If element depth is decreased, it causes an increase in resonant frequency. As the insertion of tuning elements decreases slot width, breakdown voltage is decreased, which could cause arcing. Capacitive tuning is called cookie-cutter tuning. This type of tuning arrangement consists of a metal ring that is mechanically moved. Due to mechanical and breakdown voltage problems, capacitive tuning is more suitable for lower-frequency radars. Capacitive and inductive tuning can be used on the same magnetron. With these methods, the resulting tuning range can cover up to 10% of a tube's design frequency. *Frequency agility*, the ability to rapidly change output frequency, can be obtained by using a motor-driven rotating disk or oscillating piston to alter cavity geometry in a known pattern. Frequency agility is an option that is found only in major military radar systems and is used to counter jamming by enemy forces.

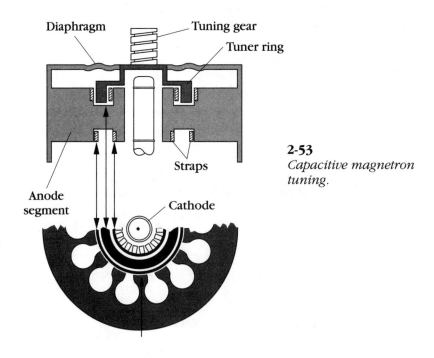

2-53
Capacitive magnetron tuning.

RF solid-state transmitting devices

Due to the rapid advances in electronics, solid-state transmitting devices are gaining in use as tube replacements, particularly in military applications. These devices have proved to be suitable with phased array antennae, such as the U.S. Navy's Aegis radars and the Army's Patriot missile control systems. Rather than having one RF transmitting component, such as a magnetron, klystron, or TWT, numerous solid-state transmitting modules are interconnected to form a high-powered transmitter. The action of interconnection is called *summed.* The idea has been studied and tried for years, but it took the rapid advances in low-cost, high-speed computer technology to allow for practical application. Through computers, the transmitted beam shape can be electronically controlled. By using four flat transmitting faces, 360-degree rotational coverage can be provided, eliminating the requirement for a heavy mechanically rotated antenna. This type of antenna is suitable for overcoming jamming and natural interference and for tracking multiple targets. An added benefit is that one phased array radar can be used for several different functions by simply altering the controlling software.

The *Gunn Oscillator* is a gallium arsenide crystal (GA) solid-state bulk-effect source of microwave energy. Construction is similar to that of conventional diodes in that it is fabricated from two blocks of semiconductor material, one P-type and one N-type. The fact that microwave energy can be generated by applying a steady voltage across a chip of N-type semiconductor material doped with gallium arsenide was first demonstrated by the researcher J.B. Gunn. RF energy emittance results from the excitation of electrons in the crystal to energy states higher than they would normally occupy. GA is an unusual material in that empty valance bands exist higher than the ones filled with electrons. What is unusual is that when electrons are excited to the point where they jump to the higher valance band, they are less mobile when influenced by an electric field. In solid-state crystals doped with different materials, excited electrons shift to higher valance bands and exhibit a high degree of mobility. Often, due to this greater mobility, they become free electrons, unattached to any atom.

Gunn oscillator operation begins with a voltage applied across a GA semiconductor. It will cause current flow to increase as the applied voltage is increased. Once the voltage is raised to a certain point, excited electrons leave their initial valance shell and rise to a higher one. So far, this is normal semiconductor operation. In the GA crystal, once in the higher valance shell, they become immobile (a curious event). If the potential is increased to a high enough level, current flow will begin to decrease as voltage is increased. If an N-type semiconductor GA crystal is doped unevenly, it will break into regions with different intensities of electric fields across them. Small domains will form within the areas under the influence of strong electric fields. Outside the domains, the effect of the electric field will be weak, with the result that the domains will be unstable.

Transmitter troubleshooting hints

When troubleshooting any equipment failure, always observe all front panel indications before opening a chassis or access panel. Corrective maintenance time required to return a defective radar system to full capability can be greatly reduced by observing all front panel controls, meters, and the video present on the displays. Many times what seems to be a problem is just a switch or control in the wrong position. For example, Fig. 2-54 is a picture of the transmitter section of an AN/ASR-8 airport surveillance radar. The meters across the panel

2-54
Radar transmitter
control panel.

monitor magnetron current, high voltage, and high-voltage power supply current. Often, by monitoring panel meter readings, you can anticipate equipment failures, decreasing down time. Another point to remember is that interruptions in input power can often trip breakers and blow fuses.

Figure 2-55 is the modulator cabinet of a linatron, an industrial X-ray system. Notice the line of breakers across the auxiliary power distribution panel. Even though the breakers control low-voltage power supplies, if one is tripped, it can still affect equipment operation. When first investigating a problem, the few minutes spent observing front panel indications can save hours in maintenance time.

2-55
Linatron modulator cabinet.

Magnetron defects are fairly easy to isolate. The most common symptom of magnetron problems is low magnetron current. If you observe the radar display, background noise or grass will appear to be normal. Any displayed echoes will be weak, possibly indistinct, and fuzzy. AFC, used to maintain magnetron frequency, might be inoperative. Testing the system with an echo box will indicate a short range.

Another problem is a weak magnet. For proper transmitter operation, it is essential that the magnetron magnet be strong and within specifications. The most common symptoms are high magnetron current and drifting AFC.

When replacing a magnetron, it must be seasoned, or baked in, for proper operation. Internal arcs are common in new tubes and are caused by the liberation of gases from the tube elements. Internal arcs can also be caused by sharp surfaces inside the tube, mode shifting, and operating at excessively high currents.

The cathode structure can withstand arcs for a short time. However, damage can result, destroying the tube. The seasoning process slowly increases the tube voltage until an arc occurs. Internal arcs during the seasoning process are caused by the tube eliminating gas. If arcing is persistent, lower the voltage. Do not resume increasing voltage until the tube settles down. Seasoning is complete when operating levels of voltage and current have been reached and the tube functions without arcing. Proper seasoning might take several hours.

Remember that radar transmitters and modulators are constructed from many high-voltage components. Problems can be avoided by attention to detail and cleanliness. Dust buildup on components can cause arcs and high resistance paths to ground. Also, dust and dirt can cause components to run hotter, leading to shorter lives and other problems.

Attention to detail is simply observing the internal condition of the equipment. Periodically check water and cooling lines for leaks, as voltage and water do not mix well. If the system is running too hot due to dirt buildup or other problems, components and cables might often be discolored. Cables can arc to the chassis and components, leaving a burnt trail or hole. I found one arced cable by the hole it burned in the Bakelite cover of a pulse transformer. Major components such as PFNs, transformers, and high-voltage capacitors are oil-filled to provide for insulation. You check for oil leaks by feeling the surface when you wipe out the inside of the high-voltage cabinets. Oil leaks indicate that catastrophic component failure is imminent. The leaky part must be replaced quickly. Also, when cleaning the internal surface of any high-voltage cabinet, use only compressed air, possibly alcohol (a cleaner that dries without a residue) and dry rags. You don't want a residue that will provide a path to ground for high voltage. Air filters must be checked and cleaned. Reduced or blocked air flow will lead to equipment failures. Cooling systems must also be periodically checked for fluid levels, biological growth, and filter cleanliness.

3

Waveguides and antennae

A very common radar component is the waveguide. By outward appearances, a waveguide is nothing more than a rectangular or circular hollow metal tube with flanges on either end so that sections can be bolted together. However, this is a crucial radar component that is manufactured to exacting standards. The inner surfaces are carefully machined smooth with a uniform cross section and are highly conductive. It is common for guides to be fabricated from brass, aluminum, or copper. Rather than carrying liquid or gas, waveguides are designed to provide a sealed path for the movement of electromagnetic energy.

Waveguide theory

There are two methods of transferring electromagnetic energy: current flow through conductive wires and the movement of an electromagnetic field. It is the latter method that waveguide theory is based on. The earliest method of transferring an electromagnetic field was the use of two parallel transmission lines. The two conductive elements, illustrated in Fig. 3-1, are capable of guiding electromagnetic fields. As shown in the drawing, the transmission lines are nothing more than uninsulated wire separated by an air gap. The function of the porcelain standoffs is to prevent the conductors from shorting out by coming into contact with any surface or each other. This simple approach is acceptable at low frequencies, but it is unusable at the much higher microwave frequencies associated with radar technology. As current flows through the conductors, the resulting electromagnetic field is perpendicular to the plane that contains the wires. At low frequencies, that is not a problem because energy losses are

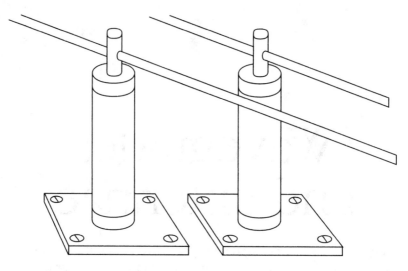

3-1 *Two-wire transmission line.*

minimal and have very little effect on operation. At microwave frequencies, RF energy escapes by radiation into free space, rendering transmission lines totally useless as conductors.

Figure 3-2 is a sketch of the electric (E) field radiation patterns encountered in a two-wire transmission line. As shown in the drawing, radiation parallel to the conductors is confined and not lost. In the perpendicular plane, the E fields produce a pattern very similar to that of an antenna. Through experimentation it was discovered that electromagnetic energy could be transferred without loss when one conductor is extended completely around the other. This should be familiar as the very common coaxial cable that is in widespread use today.

Figure 3-3 is the cross section of a representative coaxial cable. The center conductor is completely surrounded by an insulating ma-

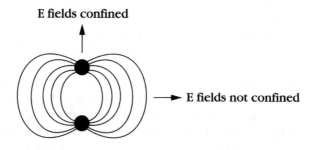

3-2 *E fields in a two-conductor wire.*

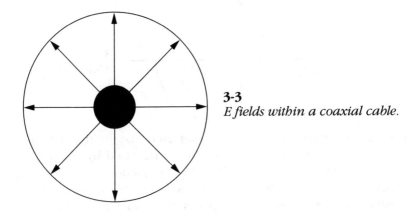

3-3
E fields within a coaxial cable.

terial, which acts as a barrier between the two conductors. The second conductor, in the form of copper shielding, is wrapped around the insulating material. A final outer layer is an insulating jacket. Selection of the outer jacket material depends on the environment that the cable is expected to be exposed to. Although coaxial cable is satisfactory for frequencies much higher than the older-design transmission line, coaxial does have inherent frequency limitations restricting its usefulness in radar applications. This limitation can be overcome by a phenomenon known as the *skin effect.* Current flow in any conductor occurs near the outer surface, hence the term skin effect. If most electron current flow is in the thin outer layer of a conductor, then the bulk of the conductive material contained within the conductor is wasted. Since that is how it functions, then why not use just the outer layer of metal as the means to transfer electromagnetic energy? A metal conductor with an empty center that propagates electromagnetic energy is called a *waveguide.*

Waveguide operation is based on the theory of quarter-wave stubs. To briefly explain, observe the sine wave in Fig. 3-4. A sine wave is one complete cycle or wave. If you begin at zero, on the left, the voltage is minimal. One-quarter wavelength along the sine wave, you have the maximum value of voltage. Continue on to one-half wave, and the voltage is once again minimum. As the cycle continues, at the three-quarter wavelength, the voltage is at its maximum negative point. From this you can conclude that the maximum transfer of energy occurs when voltage is maximum, or at the one-quarter wavelength point. The physical distance of the quarter wavelength point is completely dependent on the frequency of the applied signal.

If you refer back to Fig. 3-1, you will note the porcelain standoffs used to insulate the two-wire transmission line from its environment.

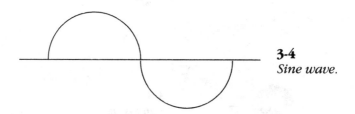

3-4
Sine wave.

Through experimentation, it was discovered that the upper frequencies' limits of transmission line could be increased by changing the
type of insulator used. Porcelain at low frequencies presents a high
impedance to ground, but at high frequencies presents a very low impedance. It was found that a superior insulator was a quarter-wave
section of transmission line shorted at one end. Figure 3-5 illustrates
the concept. Positional placement of the insulator was found not to
be crucial. The addition of more shorted quarter-wave sections increased the mechanical strength of the transmission line. If a sufficient number of sections were combined, the result was a continuous
metal pipe. In effect, the transmission lines became a part of the
walls. Essentially, a waveguide is nothing more than a continuous run
of shorted quarter-wave sections. Figure 3-6 has two views. The first
depicts the effect when additional quarter-wave sections are added
for rigidity. When enough are used to result in a solid surface, a conductive box is the final product.

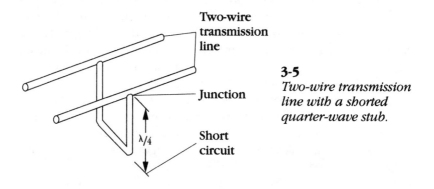

**Two-wire
transmission
line**

Junction

3-5
*Two-wire transmission
line with a shorted
quarter-wave stub.*

$\lambda/4$

**Short
circuit**

Any line or conductor that carries RF energy can encounter three
types of energy loss: copper loss, dielectric loss, and radiation loss.
Copper loss is determined by the physical resistive characteristics of
the conductor or transmission line that carries the energy. Losses are
the least in a silver conductor, and they increase if copper, aluminum,
or gold is substituted. Dielectric losses are due to heating of the insu-

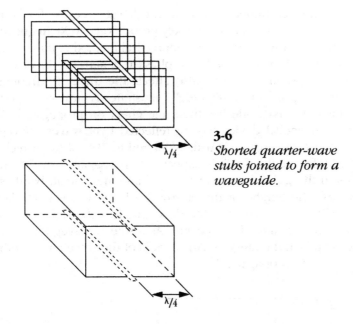

3-6
*Shorted quarter-wave
stubs joined to form a
waveguide.*

lating material between conductors in a transmission line or coaxial
cable. Radiation losses are caused by energy escaping to free space
from a conductor or line. It is now apparent that shielding can limit
or eliminate radiation losses.

When compared to other conductors, a waveguide is far superior
for carrying RF energy. It has very low copper loss due to the large
conducting surface area on the interior of a waveguide. Dielectric loss
is also very low because the guide is either pressurized with an inert
gas or air, which are excellent insulators when used with sufficient
space. Radiation loss is nonexistent, as the electromagnetic field is
completely contained within the waveguide walls. For all its advantages, it does have disadvantages that limit its use: high cost, complexity of installation, and electrical characteristics that limit its practical use
to only very high frequencies.

RF energy is propagated through a waveguide via an electromagnetic wave front. For propagation to occur, the width of the guide must
be one-half the wavelength of the energy that is to travel through it,
meaning that it is frequency limited. If the applied RF has a frequency
of 1 MHz, the correct waveguide would have to have a width of slightly
less than 500 feet! If the propagated frequency is increased to 100 MHz,
the guide width would decrease to 4.9 feet—smaller, but still not practical. With an increased frequency of 1 GHz, the waveguide width
would only have to be a modest 5.9 inches. The required waveguide

width for a given frequency is easy to calculate because for propagation to occur, guide width is inversely proportional to RF frequency. As a general rule, a waveguide is an impractical method of transferring RF energy below 100 MHz due to the physical dimensions involved.

Waveguide physical size also affects the upper RF frequencies that can be applied to it. Figure 3-7 illustrates typical waveguide dimensions. A waveguide has the same effect on propagated RF energy as an optical device has on reflected light waves. Waveguide width, the A dimension, is usually limited to 70% of the wavelength of the RF frequency that it is expected to propagate. The 70% figure results in the guide being capable of handling a small band of frequencies. The height, or the narrow B dimension, is controlled by the breakdown potential of the dielectric. It is the same concept as the working voltage of a capacitor. As a rule, a waveguide's narrow dimension is in the range of 20% to 50% of the wavelength of the RF energy it is to propagate.

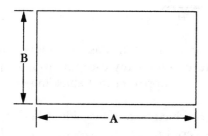

3-7
Waveguide dimensions.

RF energy propagation is dependent on both magnetic fields and electric fields. If one field is missing, then propagation is impossible. An electric field indicates a difference in potential between two points. The electric field that exists in the dielectric of a charged capacitor is the simplest form of an electric field that you will encounter. In Fig. 3-8, the positive plate of the capacitor is connected to the positive battery terminal, and the negative plate is connected to the negative terminal. The resulting positive potential on the right plate attracts electrons from the negative potential on the left plate. Arrows indicating electron motion originate on the negative plate and extend through the dielectric to the positive plate. The relative electric field strength is indicated by the density of arrows.

Figure 3-9 is the illustration of electric energy propagation in a two-wire transmission line. In the drawing, the sine wave depicts the propagation of RF energy. On the positive alternation, the resulting E field arrows point up, indicating that the bottom of the guide is at

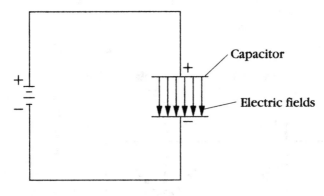

3-8 *E field in a capacitor.*

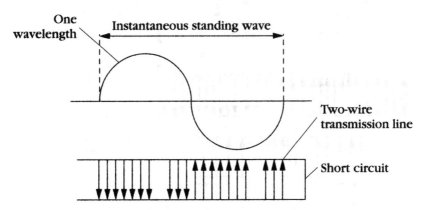

3-9 *E fields in a waveguide compared to the applied RF.*

a negative potential and the top is at a positive potential. On the negative alternation, the arrows point down, indicating that the field has changed polarity. As with the capacitor drawing in Fig. 3-8, the number of arrows indicates the field density. The electric field in a waveguide is called an *E field*.

E field motion is very similar in a waveguide, as illustrated in Fig. 3-10. It is a drawing of the E field that results across a one wavelength section of waveguide. The three views are from the top, side, and end of the waveguide. As with the previous drawings, E field strength is proportional to the density of arrows. At time 0, applied voltage is zero, resulting in no E field. As the sine wave increases in amplitude, the arrow density increases toward maximum. When the standing voltage is at its maximum positive peak, the E field is at its peak intensity. As the sine wave decreases, arrow density decreases. Both the

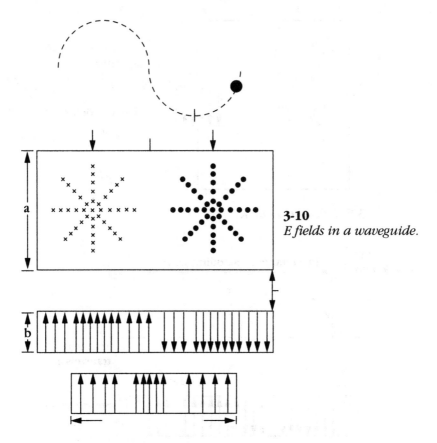

3-10
E fields in a waveguide.

applied voltage and field strength are minimum at the same point in time. Field polarity reverses as the sine wave decreases below zero and increases in negative value. After passing through the maximum negative value, both the E field strength and sine wave increase toward the positive. This cycle continually repeats itself. In view C, notice that field intensity is greatest at the center of the waveguide. As you move toward either wall, field strength decreases to zero.

The magnetic field felt in a waveguide is caused by the current flow through the conductive material coating the interior walls. The presence of a magnetic line of force is illustrated by a closed loop. For a magnetic line of force to exist, it must be a closed loop. All lines associated with current flow are called a magnetic, or H, field. H field strength is directly proportional to the magnitude of current flow. The direction of H lines is determined by the left-hand rule, illustrated in Fig. 3-11. As shown, the conductor is grasped with the left hand. The extended thumb points in the direction of current flow. All other fingers are wrapped around the conductor point, and they point in the

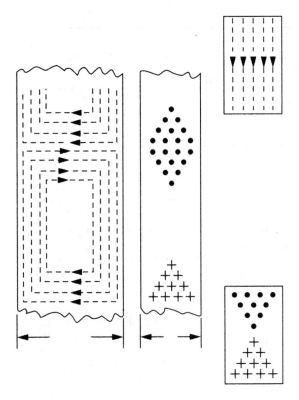

3-11 *H fields in a waveguide.*

direction of the resulting H lines. E and H fields exist simultaneously in a waveguide. Current flow causes the E field, which in turn causes the magnetic H field.

Figure 3-12 depicts a magnetic field in a waveguide resulting from the presence of an E field. Once more, field density is indicated by the line density. As was previously stated, it is impossible for an open-loop magnetic field to exist. The drawing has three views: end, top, and side. From the figure, it is evident that the magnetic field is strongest near the waveguide walls and the weakest in the center. That is because H-field strength is directly proportional to the E-field strength, and electron current flow is along the waveguide walls. Another condition is that the magnetic field is parallel to the waveguide surface.

Boundary conditions are similar to the propagation of RF energy in free space. With a waveguide, all the RF energy is contained within the guide structure. Two conditions must be met for RF propagation to occur in a waveguide. First, the E field must be perpendicular to a conductive surface. Secondly, there must be no magnetic field com-

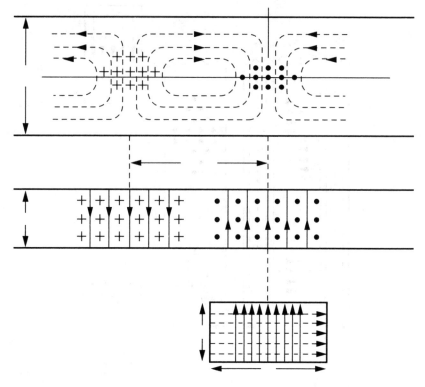

3-12 *E and H fields in a waveguide.*

ponents perpendicular to the waveguide surface. That condition is met because magnetic field lines are parallel to the surfaces.

Modes of operation

The dominant mode of operation results from the physical characteristics of a particular waveguide. It is the most efficient mode of operation because it has the least attenuation on the RF energy. The lowest frequency that a guide can propagate is determined by the "a" dimension. Two fields simultaneously exist in a guide: transverse E and transverse M. In the TE mode of operation, all of the E field is perpendicular to the length of a guide. It is impossible for any E-field components to move parallel to the direction of wave front travel. In the TM mode, the H field is perpendicular to the length of the guide. Once again, as the H field must be perpendicular, no H field component is parallel to the direction of wave front travel. A subscript number is used to describe the field pattern in a given mode of operation.

The first subscript indicates the number of half wave patterns in the "a" dimension. The second subscript indicates the number of half wave patterns in the "b" dimension of the guide.

For example, look at Fig. 3-13. The illustration is of a rectangular waveguide. The E field lines are perpendicular to the direction of wave front movement. This is classified as TE mode of operation. In the wide dimension, E field varies from 0 at the wall surface to maximum in the center of the waveguide. The observed patterns result in a 1 subscript because there is only 1 complete half cycle. In the narrow dimension, the E field is parallel, so there is no intensity change, which results in a second subscript of 0.

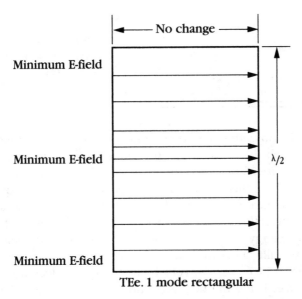

TEe. 1 mode rectangular

3-13 *Dominant mode of operation in a rectangular waveguide.*

As all circular waveguides have the E field lines perpendicular, they operate in a TE mode of operation. The circumference determines the first subscript. The second subscript is determined by the diameter of the waveguide.

Electromagnetic action in a waveguide

For RF energy to be propagated through a waveguide, it must be coupled in and out of it. RF energy can be transferred into or out of a waveguide through the use of a probe, which acts as an antenna. When it is used to input energy, an alternating waveform is radiated from the probe, just like a transmitting antenna. Figure 3-14 depicts a probe inserted into a waveguide as a means of coupling energy. The RF energy that is radiated along line "B" is attenuated because it doesn't meet boundary condition requirements. RF energy radiated along lines A and C will form into a wave front as it meets boundary conditions. The resulting wave front is then propagated down the guide.

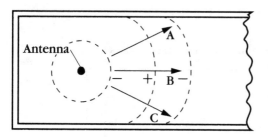

3-14
Probe inserted into a waveguide to couple in RF energy.

RF radiation within a guide is similar to optical phenomena. The angle at which the wave front strikes the waveguide wall is called the *angle of incidence*. The angle that the wave front is reflected from the wall is called the *angle of reflection*. As in basic optics, the angle of incidence must equal the angle of reflection. The angles of incidence and reflection are a function of the frequency of the propagated RF energy. The effects of frequency on propagation in a waveguide is illustrated in Fig. 3-15. As frequency increases, the resulting angles increase. Conversely, as frequency decreases, the angles decrease. If the frequency continues to decrease, the resulting angles decrease to zero, and propagation ceases. That frequency is called the *cutoff frequency*. At that point, RF energy forms standing waves, which are reflected perpendicular to the waveguide walls. This results in no energy being conducted down the waveguide.

Wave motion in a waveguide appears to be unusual. Relative motion in a waveguide is illustrated in Fig. 3-15. The propagated wave front appears to move at the velocity of light as it is reflected along a waveguide. Actually, as it is propagated, its velocity is less than that of light. The straight line, or axial motion, is called the *group velocity*. The relationship between group and diagonal velocity is unusual because in the time that the wave front moves from point one to point two, it is at the speed of light. The diagonal movement, due to reflecting off of the walls, moves a shorter distance. If the wave front

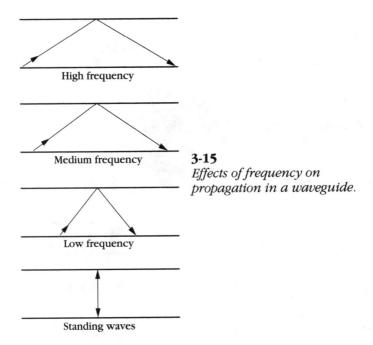

3-15
Effects of frequency on propagation in a waveguide.

position is observed with instrumentation, the wave front appears to actually move faster than the speed of light!

The phase of the propagated RF will change as it moves along a waveguide, and this is called the *phase velocity*. If a modulated RF signal is propagated down a waveguide, the modulation envelope appears to move at the group velocity. The individual cycles within the envelope would appear to move through it at the phase velocity. If frequency decreases, group velocity decreases, and phase velocity increases. Group velocity will always be less than that of light, by the amount that the phase velocity appears to be faster. The concept is illustrated in Fig. 3-16. P is the phase velocity of the propagated RF. G is the group velocity. Accurate measurement of the actual event requires the use of sophisticated laboratory instruments.

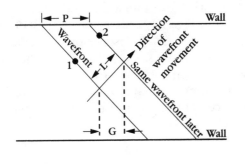

3-16
RF propagation in a waveguide.

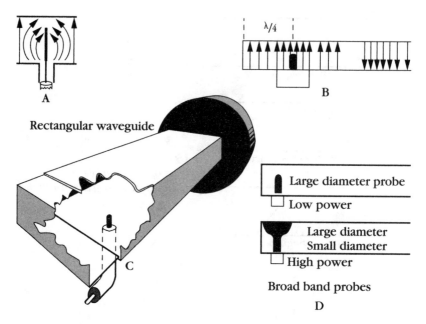

3-17 *Probe coupling RF energy into and out of a waveguide.*

There are three methods of coupling RF energy in and out of a waveguide: probe, loop, and aperture. Probe, or capacitive coupling, is the same as a conventional ¼-wave antenna and is illustrated in Fig. 3-17. The probe must be located in the center of the "a" dimension, ¼ wave length, or an odd multiple from the short circuit end of the guide. Maximum coupling occurs when the probe is in the center of the maximum E field. The degree of coupling is varied by the depth of the probe in the waveguide. Loop coupling, presented in Fig. 3-18, is inductive coupling whereby the loop is placed at the point of maximum H field intensity. Current flow in the loop induces a magnetic field within the waveguide, and the resulting degree of coupling can be varied by rotating the loop. The final method is aperture, or *slot coupling*, and that is depicted in Fig. 3-19. Notice that Fig. 3-19 has several slots in the waveguide. A waveguide installation will typically have only one slot per application. If the slot is positioned in the area of maximum E-field strength, it is then known as *electric coupling*. By positioning the slot in the area of maximum H field, it is then called *magnetic coupling*. If it is placed at the point of maximum E and H, it is *electromagnetic coupling*.

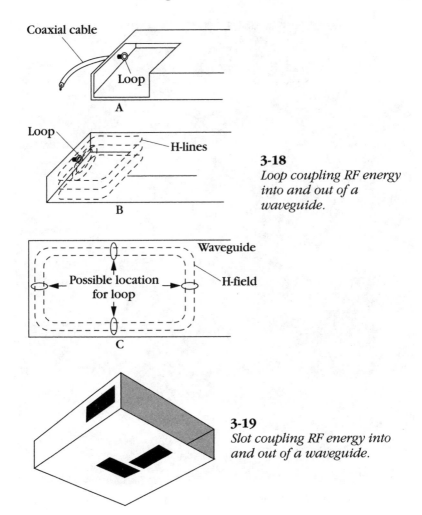

3-18
*Loop coupling RF energy
into and out of a
waveguide.*

3-19
*Slot coupling RF energy into
and out of a waveguide.*

Waveguide installation

An ideal waveguide installation would be just one continuous, straight piece from transmitter to antenna. Any abrupt changes in size, shape, joints, or direction have the potential of causing reflections, which results in loss of power and can translate into reduced range and resolution. Far from the ideal straight run, many installations require bends, twists, and multiple pieces of waveguide to connect the antenna, transmitter, and receiver. Figure 3-20 is a photograph of the RF portion of an AN/SPS-48E long-range, three-dimensional air-search

3-20 *Waveguide installation in a radar system.* ITT Gilfillan, a unit of ITT Defense and Electronics

radar. The waveguide is the light-colored rectangular pipe in the upper center of the picture. Notice that there appear to be three bends in this short section of guide. Also, the individual pieces of guide are flanged and connected together with bolts. The placement of the bands and size of the flanges are not random.

A bend is acceptable as long as it has a radius of at least two complete wavelengths of the propagated RF energy. Figure 3-21 is a conventional bend. To minimize reflections, it is very gradual and presents no sudden changes to the propagated wave front. Figure 3-22 is used if a sharp 90-degree bend is required. To prevent reflections, two 45-degree bends, ¼ wavelength apart, are used. The use of

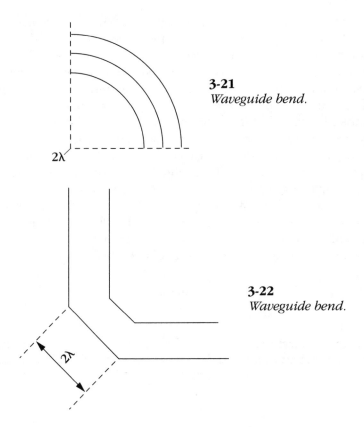

3-21
Waveguide bend.

3-22
Waveguide bend.

two bends minimizes reflections, as one bend will reflect directly and the other will reflect a mirror image. Any reflections will electrically cancel one another out, as if they did not exist in the first place, with the result that only the desired RF signal is left. Twists are far more common than one would imagine in radar installations. Any twists also require two wavelengths to prevent reflections. The concept is illustrated in Fig. 3-23.

When two sections of waveguide are joined together, any imperfections can result in reflections and subsequent power losses. The highest-quality joint is fabricated during the manufacturing process.

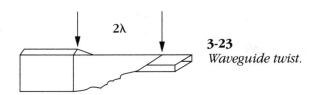

2λ

3-23
Waveguide twist.

The sections are machined to very exacting tolerances and permanently joined by welding that results in a smooth, hermetically sealed joint. Most other joints, due to installation requirements or to provide access for maintenance requirements, are designed to be semipermanent.

Semipermanent joints must be very carefully designed. Figure 3-24 illustrates the physical design of a semipermanent joint. In this example, the flange on the left has a flat surface. The right side has slots ¼-wavelength deep on the inner surface, located ¼ wavelength from the point where the flanges join together. The ¼ wavelength is important because electrically it represents a short circuit. The two ¼-wavelength slots join to form a ½ wavelength section. The ½-wave section reflects the short circuit to the point where the waveguide walls from the two sections join together. This type of joint results in an electrical short circuit being created at the junction point, eliminating any possibility of a reflection being formed.

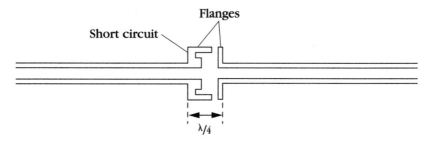

3-24 *Waveguide joint.*

Figure 3-25 is a representation of the inner surface of the waveguide joint. Notice that the waveguide dimensions are located ¼ wavelength from the short-circuit channel. Due to the phenomenon associated with the actions of RF in waveguides, the sections might be separated by as

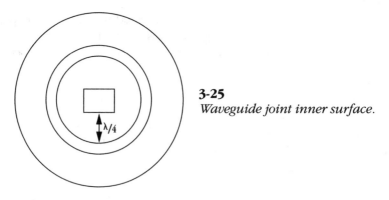

3-25
Waveguide joint inner surface.

much as ⅒ of a wavelength and still function without an excessive loss of energy. The separation is useful because it allows for the insertion of gasket material to ensure pressurization of the waveguide. Pressurization is vitally important because it prevents moisture from entering the guide and causing electrical and corrosion problems. Permanent manufacturer-fabricated joints experience losses in the range of .03 dB, and semipermanent joints experience losses of .05 dB.

Rotating joints, very important to rotational antennae, have special design consideration. A circular waveguide is normally found in the connection between a rotating antenna system and the transmitter-receiver section. It is ideal for this application because the waveguide rotates with the antenna, while the electromagnetic fields that propagate the RF energy are fixed. The most common configuration uses a circular rotating section, with fixed rectangular sections on each end to facilitate connection with the antenna and the remainder of the system. Figure 3-26 shows a typical rotating RF joint. This joint

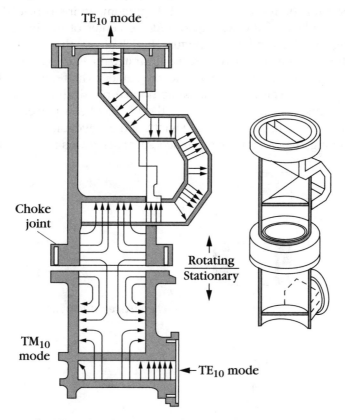

3-26 *Typical radar rotating joint.*

is similar in construction and concept to the fixed joint. The same dimensions and physical design are used to minimize reflections.

Impedance matching is crucial in radar systems to ensure minimum losses of RF energy. The RF transmitting system must be perfectly matched with the load, or antenna. A mismatch results in standing waves and reflections, which in turn causes loss in power. Power losses equally affect the transmitted energy and the received echoes. As an impedance mismatch reflects energy, it decreases the power-handling capability of the affected waveguide. Impedance-changing devices are used to induce a desirable standing-wave ratio (SWR) of 1:1. The optimum value of load impedance ensures that the source transfers full power to the source. It is similar in concept and effect to inductors and capacitors in conventional circuits.

Waveguides cannot be terminated in a resistance, as a conventional circuit would be. Devices used in waveguides for impedance matching are called *irises*. Irises are used to insert the required capacitance or inductance to match the waveguide to the remainder of the system. Little more than a metal diaphragm, an iris is placed in a transverse plane of the waveguide. At the center of the irises is an opening called a window or aperture, which allows the RF energy to pass through. An inductive iris, as illustrated in Fig. 3-27, places a shunt inductive reactance across the waveguide. The figure has three different views. The upper one is the normal physical dimensions of the waveguide. In the center view, the iris has been placed on the narrow dimension, decreasing the width of the waveguide. The wider the opening, the greater the reactance. A capacitive iris places a shunt

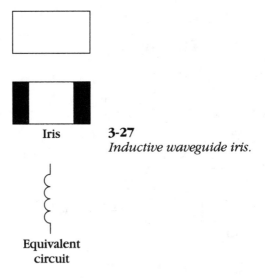

Iris

3-27
Inductive waveguide iris.

Equivalent
circuit

capacitive reactance across the waveguide and is depicted in Fig. 3-28. As with the inductive iris, the wider the opening, the greater the capacitive reactance. The difference in this installation is that the iris is placed along the wide dimension of the waveguide. A third type of iris is pictured in Fig. 3-29. Electrically it functions the same as a parallel inductive-capacitive reactance. At resonance, it places a high shunt reactance across the line. At frequencies above or below the RF, it acts as either an inductive reactance or capacitive reactance circuit.

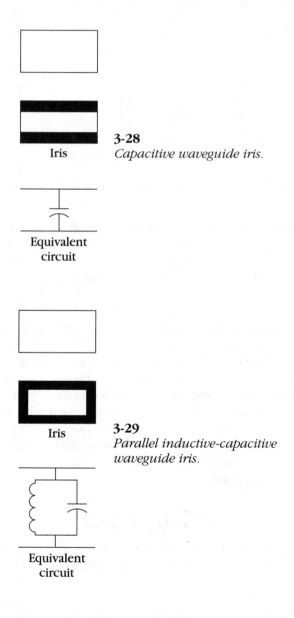

Iris

3-28
Capacitive waveguide iris.

Equivalent
circuit

Iris

3-29
*Parallel inductive-capacitive
waveguide iris.*

Equivalent
circuit

Posts and screws can also be used as impedance-matching devices in a waveguide installation, as illustrated in Fig. 3-30. Partial penetration with the screw induces a shunt capacitive reactance into the guide. If it makes contact with the top and bottom walls, it acts as an inductive reactance.

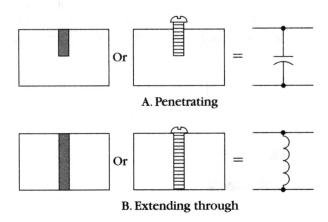

A. Penetrating

B. Extending through

3-30 *Waveguide impedance matching.*

Because a waveguide is frequency-sensitive, termination is vital to system operation. Electrically, a waveguide acts as a single conductor, which makes it difficult to define its characteristic impedance. The characteristic impedance of a waveguide is approximately equal to the ratio of the E field strength to the H field strength. It can also be said that it is equivalent to the voltage-to-current ratio in a coaxial cable without standing waves. Termination can be accomplished in one of several ways, as illustrated in Fig. 3-31. Graphite sand or a resistive rod can be used to terminate and absorb excess energy. If it is desirable to reflect the energy back from the terminated end, then a metal piece can be permanently installed across the waveguide. To install a removable termination and still prevent reflections, an end cap with ¼-wave cavities is installed. By placing at a point of minimum current flow, the E field is not attenuated. The final method is an adjustable plunger. It is very similar in concept and construction to a rotating choke joint.

Duplexer

The vast majority of radar employs the same antenna system for both transmit and receive functions. When two such dissimilar functions

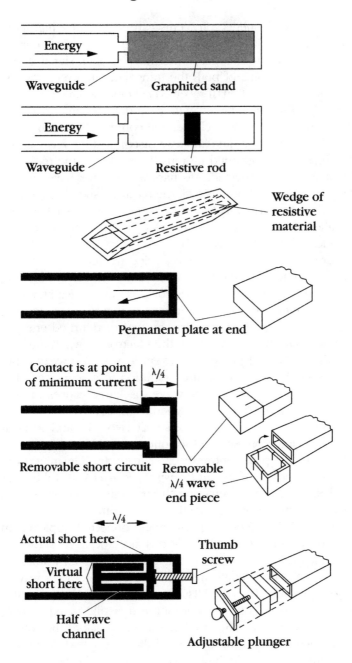

3-31 *Waveguide termination techniques.*

use the same components, problems can arise. If the high-powered output of the transmitter is applied directly to the receiver, damage results. Weak echoes from target do not have enough signal strength to be routed to the inputs of both the transmitter and receiver and still provide usable information. If received echoes are directed to both the transmitter and receiver, targets will go undetected due to signal loss. The solution is to have a switch that will connect the transmitter to the antenna during transmit time and also disconnect the receiver. During receive time, all the returned RF must be routed to the receiver, with the transmitter disconnected. Mechanical switches have the power-handling capabilities, but they lack the required speed. After much research, the conclusion was found to be electronic switches. Electron tubes, called *transmit-receive tubes* (TRs), are mounted in a section of waveguide called a *duplexer*.

A duplexer has two functions. During transmit time, it connects the transmitter to the antenna and isolates the receiver to prevent damage. After the main bang, it disconnects the transmitter and connects the receiver to the antenna. It must switch between transmit and receive function rapidly to prevent the loss of returned energy from close targets. During its operation, the electronic switch must absorb a minimum amount of power to prevent losses from weak echoes. A TR tube is similar in concept and operation to a spark gap. The application of a high-current microwave discharge causes the tube to complete a low-impedance path when it ionizes.

A conventional duplexer consists of two TRs and a mounting waveguide to interconnect the transmitter, receiver, and antenna system. The TR tube protects the receiver from the transmitted RF. The second tube, known as an *antitransmit-receive* (ATR) tube, prevents returned RF echoes from entering the transmitter. Duplexers can be connected in either series or parallel configurations.

Figure 3-32 illustrates a typical series-connected duplexer. In duplexer construction, physical dimensions are crucial. As shown in Fig. 3-32, the TR spark gap is located ½ wavelength from the T junction. The ATR is placed ½ wavelength from the waveguide, and ¾ wavelength from the T junction. The two TRs are separated by ¼ wavelength. During transmitter time, both tubes are ionized, as illustrated in Fig. 3-33. The action of both tubes firing reflects a short back to the waveguide, removing the receiver from the path of the transmitted RF. Figure 3-34 shows circuit operation during receive time. When the receiver is online and active, both tubes are not ionized, due to the low-level RF returned from the targets. That has the effect of reflecting a short to the point just before the receiver, disconnecting the transmitter from the antenna. That, in turn, channels all the returned RF to the receiver.

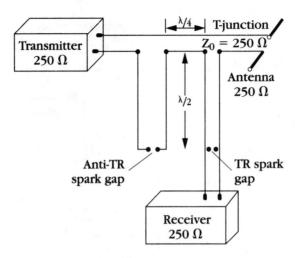

3-32 *Series configured duplexer.*

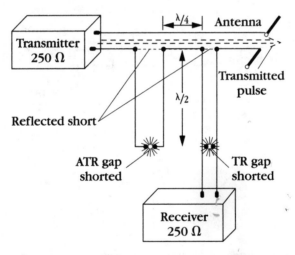

3-33 *Series configured duplexer during transmit time.*

The parallel configuration, depicted in Fig. 3-35, appears very similar at first. The major difference is in the physical placement of the TRs. The TR is installed ¼ wavelength from the T junction. The ATR tube is located ¼ wavelength from the T junction in a shorted stub. The stub is ½ wavelength long. The result is that electrically the TR and ATR tubes are mounted ¼ wavelength apart.

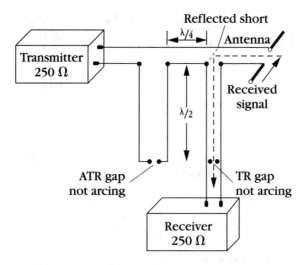

3-34 *Series configured duplexer during receive time.*

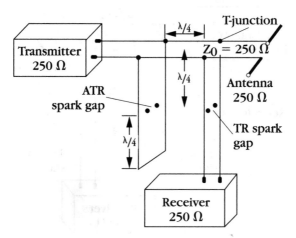

3-35 *Parallel configured duplexer.*

Figure 3-36 illustrates circuit actions when the transmitter fires. During transmit time, both tubes are ionized by the high-powered RF pulse. By ionizing, both tubes reflect open paths to the RF in the waveguide. All transmitter output power is routed to the antenna system. Figure 3-37 shows circuit conditions during receiver operation. Again, during receive time, the low-powered echoes do not ionize the two tubes. That has the effect of reflecting an open to the point just before the receiver branch. All the returned energy is routed to the receiver for processing.

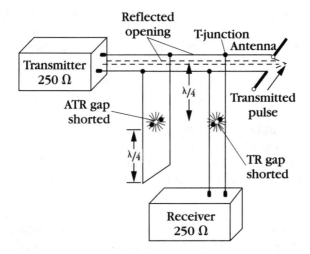

3-36 *Parallel configured duplexer during transmit time.*

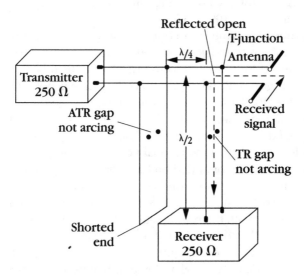

3-37 *Parallel configured duplexer during receive time.*

TR tube construction can range from a device as simple as two electrodes across the waveguide to a glass-enclosed and evacuated electronic tube. In the more complex design, a TR tube is a conventional spark gap enclosed in a partially evacuated, sealed glass envelope. Prior to arcing and ionization, it displays a high impedance. The output pulse causes the tube to arc, changing it to a low-impedance path. The low impedance presents an electrical open at the input of the radar receiver, blocking the transmitted RF pulse. At the end of transmit time, the arc is rapidly extinguished to reconnect the receiver to the antenna. The time it takes the TR to extinguish the arc and reconnect the receiver is recovery time. The arc is formed as electrons are conducted through the ionized gas. The voltage required to ionize the gas can be decreased if the internal pressure of the tube is decreased. Recovery time can be reduced by inducing a water vapor in the envelope (check). Typical TR recovery time is in the range of 0.5 microsecond.

It is easy to see how recovery time can affect the reception of close targets (0.5 microsecond translates to 80 yards). If, due to age or other problems, recovery time increases to 1 microsecond, any targets within 160 yards of the radar would be invisible. This change in distance would be crucial if the radar is used as a shipboard navigation system. Such changes can and do happen. The time it takes the tube to ionize and protect the receiver is another crucial characteristic and is called the *breakdown speed*. In many designs, it is increased by priming the tube with a keep-alive voltage. Keep-alive voltages typically range from 100 to 1000 volts.

ATR tubes are similar in concept and design. The primary function of the tube is to block the output of the transmitter to ensure that all the received RF is routed to the receiver. This tube is pressurized with an inert gas, typically argon. Recovery time is not a factor; therefore the tube does not require a keep-alive voltage.

Tube failure is usually caused by a gradual buildup of metal particles dislodged from the electrodes. The particles act as a conducting area that lowers the Q of the resonant cavity, causing power loss. If the tube is not replaced, continued buildup forms a detuning wall, and the tube stops functioning. A second common failure is when the gas can be absorbed by the metal electrodes. Absorption causes internal pressure to decrease, with the result that breakdown or ionization becomes difficult and finally stops. Both failures cause transmitted RF to be coupled into the receiver, eventually causing damage. A good tube has a clear glass envelope. When deterioration begins, the particles discolor the envelope. Part of your preventive maintenance should include inspecting the glass envelopes to ensure that they appear normal.

As with other areas of radar technology, antenna switching is also moving forward. New duplexer hardware includes a device that can be known as the four-port circulator or the hybrid ring duplexer, and it is based heavily on waveguide theory.

The hybrid duplexer, depicted in Fig. 3-38, is fabricated in a ring shape and has four exit-entrance ports. Operation is presented in terms of the E fields. During transmit time, the RF enters arm 3 and

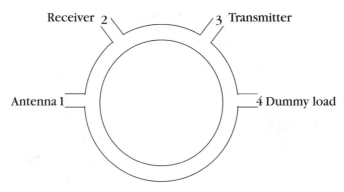

3-38 *Hybrid ring configuration.*

splits into fields, 180 degrees out of phase. One field moves clockwise, the other counterclockwise. For RF energy to be propagated down a given arm, the two fields must be 180 degrees out of phase at the entrance. The field moving in a clockwise direction ionizes the TR tube mounted in arm 4. The action of the ionized TR tube reflects an open circuit back to the ring, blocking any RF from being propagated down the arm. At this point, both fields are out of phase, but the ionized TR tube blocks the entrance, protecting the receiver from the high-energy pulse. The clockwise field ionizes the TR tube mounted in arm 2, reflecting a short back to the ring. At this point in the ring, the two rotating fields are in phase, eliminating the possibility of any energy entering the arm. The two fields continue around the ring, meeting at arm 1. At this point, the fields are out of phase, and a TR tube is not mounted in this arm. As the path is clear, the high-powered RF pulse is propagated out of the duplexer to the antenna system.

The receive function works under the same conditions. At the antenna arm, the received RF splits into two fields. The very weak received fields lack the potential to fire the TRs mounted in arms 2 and 4. As both fields arrive at phase only at arms 2 and 4, energy is propagated down both of them.

Similar in concept is the four-port circulator, illustrated in Fig. 3-39. All ports are positioned one-quarter wavelength apart. When the transmitter fires, RF energy circles in a clockwise direction. As the antenna is located exactly one-quarter wavelength away, maximum energy is transferred out that port. If you continue around the circulator, the receiver is the next port. As it is located one-half wavelength away from the transmitter, no power is passed into that port. During receive time, energy is directed in a clockwise direction to the receiver. As with the hybrid ring, TRs and ATRs are used to ensure that all energy is directed in the proper direction at the required time.

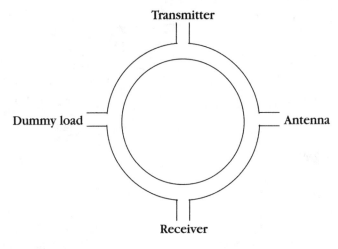

3-39 *Four-port circulator.*

Radar antennae

The antenna system is crucial to proper system operation. Table 3-1 lists characteristics that are important to all radar antennae. Mechanical considerations include size and weight. While not as crucial in a shore-based system, they are extremely important in shipboard installations. The size and weight of a radar antenna affects how a ship rides and reacts to winds and the wave motion of the surface of the ocean. That is because a large, heavy antenna mounted on a massive supporting structure raises the vessel's center of gravity, resulting in more pitching and rolling action.

Table 3-1
Antenna characteristics

Electrical	Mechanical
Frequency	Physical size
Gain	Weight
Polarization	Reliability
Beamwidth	Maintainability
Beam shape	Wind load
Side lobes	Shock resistance
VSWR	Vibration resistance
Power rating	Environmental limits

Reliability and ease of maintenance are important in today's climate of reduced funding. Often, shipboard antenna systems are stabilized to compensate for pitch and roll induced by the ocean. Other factors such as weather susceptibility determine how harsh environments will affect its operation. Electrical considerations are also important to remember. The antenna must be matched to the impedance, power output, and frequency of the RF generator. Gain, beam width, and side lobes have a direct impact on the overall operational efficiency.

Radio waves react very similarly to light waves in that they travel in straight lines and are essentially line-of-sight. Also, radio waves can be reflected, focused, and concentrated to improve effect. Radar antennae are crucial to system performance, as illustrated in Fig. 3-40. An antenna's function is to concentrate energy into directional beams capable of illuminating areas of interest. It must concentrate transmitted energy into a narrow beam to provide for the maximum power in the smallest possible beam width to illuminate small targets. The antenna system must also concentrate weak echoes received from targets to levels usable by the receiver. An antenna must match the impedance of the waveguide to that of free space to provide for maximum transfer of energy. Rotational speed and angular position must be accurate for determining the precise location of targets. An antenna that has identical transmit and receive beam patterns is called a *reciprocal antenna*. If patterns differ, then it is called *nonreciprocal*.

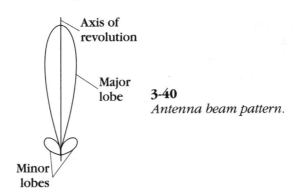

3-40
Antenna beam pattern.

Directional antennae

Important antenna characteristics include directivity, power gain, beam width, and angular accuracy. As previously stated, the most important feature of a radar system is the ability to accurately determine the position of objects. The antenna system is the function that determines how well a particular system can provide that data.

Antenna directivity is the degree of sharpness of an antenna beam. The higher the directivity, the sharper the resulting beam, and the more concentrated the radiated power. By increasing the directivity, the result is a narrower beam, which translates into less power required to cover the same area and a greater chance of detecting targets. If a radiated beam is narrow in one plane, it then has a high degree of directivity in that plane. Conversely, if it is broad in a plane, then that plane has low directivity. For example, if an antenna has a narrow vertical beam and a wide horizontal beam, then the vertical plane has a high directivity, and the horizontal plane has low directivity.

Beam width is another important antenna characteristic. *Horizontal beam width* is the width of the resulting beam in the horizontal plane measured in degrees. *Vertical beam width* is the width of the resulting beam in the vertical plane also measured in degrees.

Antenna power gain is the ratio of radiated power to that of a known reference, a basic dipole. For accuracy, both antennae must be identically fed under the same circumstances for validity. Due to the nature of radar, antennae must have a power gain for system operation. That is because although a system might radiate a pulse measured in the megawatt range, returned echoes are measured in microwatts.

To accurately place a detected object, its position is stated as a range and an angular position, as illustrated in Fig. 3-41. The angular position is measured from a known reference point, which can be ei-

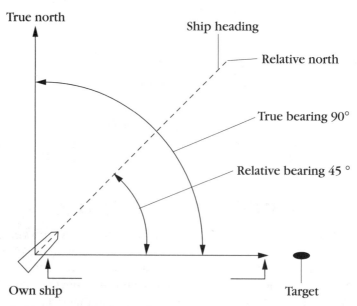

3-41 *Antenna relationships.*

ther true north or relative north. Relative north is when the bow of a ship or aircraft is used as the reference point. A rotating antenna beam provides the simplest method of measuring angular azimuth information and displaying it. Altitude data is obtained by using a steerable antenna, which actually provides slant range to target. The resulting antenna elevation data is converted to an angular position, which is then compared to a known reference—the earth's surface. Using trigonometry, radar circuits calculate target altitude from the known slant range and angle of antenna beam. Elevation angle is the number of degrees between the axis of the radar beam and the surface of the earth.

Radar antenna beam patterns

Radiated RF energy from an antenna forms a pattern. Different radar applications will require different radar beam patterns. Beam pattern is determined by the physical construction of the antenna radiating element.

Air traffic control and long early-warning radar need to detect very small targets at long range. To provide the necessary coverage, they are designed to use transmitted high power with a wide beam. Because detection ranges are 200 to 300 miles, target positional accuracy is a secondary concern. Tracking, precision approach, and weapons guidance radar require a very narrow beam because positional accuracy is the paramount requirement. Additionally, these types of systems usually must provide three-dimensional data-range, bearing, and altitude. The signal strength of the returned echo is a function of the angular position of the target within the lobe, or beam pattern, so the antenna must be capable of being steered by the operator. With all radar, the greatest target strength is obtained when the axis of the lobe passes through the target. When the antenna is rotating, as with a search radar, this fact is not noticed because the beam is constantly moving. With racking radar, as the antenna is movable, the operator knows when the beam axis is passing through a target, as signal strength is viewed on a display.

Ideally, a transmitted pulse consists of a single beam of RF energy called a *lobe*. In reality, the pattern will consist of one main lobe and several smaller side lobes. It is desirable to minimize side lobes, as it will result in a higher directivity of the beam, reduction in susceptibility from interference, and reduction of possible interference to other radar. Suppression is important because it eliminates or reduces the possibility of detecting a target in a side lobe. That is an undesirable event, as it would provide erroneous targets that could lead to inaccuracies and accidents. The existence of minor side lobes also results in a loss in output power, which reduces the system's ability to detect targets.

An example of the effectiveness of reducing side lobes is illustrated by the U.S. Navy's Aegis air defense system. The phased arrays were originally designed to detect aircraft and low-flying missiles. During the Gulf War, it was noted that some systems could detect high-flying Iraqi scuds. As the system uses computers to form the transmitted beam, by a simple and cheap software change, the system's effectiveness was improved to such a degree that it was capable of detecting and engaging scuds shortly after liftoff. By reducing the side lobes, the effective transmitted power was almost doubled without expensive hardware modifications.

Currently there are two common methods of converting radar energy to a form that will provide directivity: linear arrays and quasi-optical. Although there are several different types of radar antennae, the most common in use today is the parabolic reflector, which would be classified as a quasi-optical antenna. Radiated energy from a conventional communications antenna spreads in a spherical pattern as it leaves a dipole radiating element. A spherical pattern, although useful for omnidirectional antennae, is not suitable for forming a pencil-sharp beam of RF energy. As the energy travels farther from the antenna, it produces a sphere-shaped pattern. For radar purposes, a plane wave front is desired. In a plane wave front, the RF moves forward in the same direction and phase. To convert a spherical pattern to a plane wave front, the optical properties of microwave energy are used. As propagated RF is line-of-sight, the antenna must concentrate the energy into a plane wave front beam.

A parabolic reflector radiates RF energy in a similar fashion to the way that light waves move. The parabolic reflector is called quasi-optical, in recognition of the similarities. In Fig. 3-42, the focus represents the antenna's feed horn. The feed horn radiates RF energy developed by the transmitter to the parabolic reflector, and it is reflected parallel to the axis. By design, the focus is perpendicular to the axis. All lines indicating radiated RF are parallel to the axis, and are the same length due to reflection. The spherical RF wave front leaving the focus is converted into a plane wave front, and all lines are in phase. In effect, the reflector forms the RF energy into the pencil beam. Returned energy from targets is focused by the reflector and collected by the feed horn. Received energy that is in phase and parallel to the axis is additive at the focus. Any received RF that is not parallel to the axis or in phase is not additive, and it is ignored.

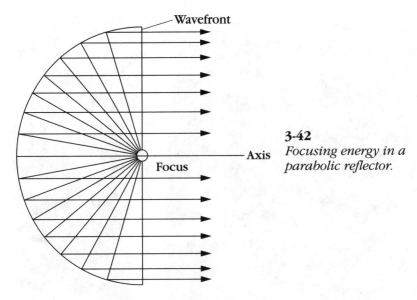

3-42
Focusing energy in a parabolic reflector.

Figure 3-43 depicts several different types of radar antennae currently in use. When a parabola is capable of rotation about its axis, then it is classified as a paraboloid, of which there are many types in current use. An unmodified parabolic reflector is the type used in air-

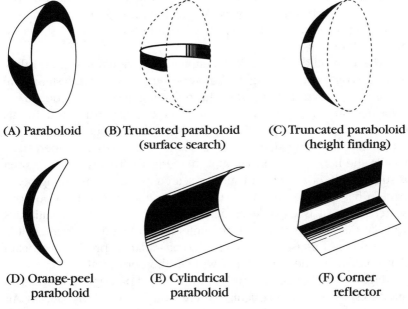

(A) Paraboloid

(B) Truncated paraboloid
(surface search)

(C) Truncated paraboloid
(height finding)

(D) Orange-peel
paraboloid

(E) Cylindrical
paraboloid

(F) Corner
reflector

3-43 *Typical radar reflectors.*

3-44 *Airborne radar system with antenna installed.* Historical Electronics Museum

borne and weapons-control radar. Figure 3-44 is a photograph of an airborne radar system. The antenna system is the green dish on the left side of the equipment. By examining the photograph, you can see how compressed a radar system has to be in order to be installed in an aircraft or missile.

The truncated paraboloid antennae are very common, and they are the type usually thought of as representing a radar system. The antenna is drawn in Fig. 3-43. The reflecting surface is reduced, or truncated, in one plane. The vertical plane is cut, or truncated. In this design, in the vertical plane the beam spreads out, but it is narrow in the horizontal plane. The resulting beam pattern is fan shaped (narrow in the horizontal and wide in the vertical). This type of antenna is suitable for air-search and surface-search radar. The resulting beam pattern has a major advantage that is desirable in a general-purpose search radar: it is capable of detecting aircraft at different altitudes without tilting the antenna. If the antenna is rotated 90 degrees, it is now suitable for use with height-finding radar. It produces a beam that's narrow in the vertical and wide in the horizontal.

Figure 3-45 is a photograph of a truncated paraboloid used as an air-search radar. This particular installation is an AN/SPS-58, an Air Force air-search radar. This particular view is from the rear, which allows you to see the waveguide, the rectangular shape in the lower

3-45 *Typical radar antenna.* Historical Electronics Museum

right-hand side of the picture. Another good view of a truncated paraboloid is Fig. 3-46. This view is from the front and gives an excellent view of the feed-horn system. This particular antenna is installed
on an AN/ASR-8 airport surveillance radar system.

Several air traffic control radar are designed to use two truncated
paraboloids for air traffic control purposes. Known as Quads, one antenna is mounted to provide azimuth data, the other elevation. Figure 3-47 is a photograph of such a system, the shipboard AN/SPN-35.
In search modes of operation, the azimuth antenna is active, providing 360-degree coverage. When precision radar information is required to provide landing instructions to an aircraft, the unit is
switched into that mode of operation. At that time, both antennae are
engaged, sweeping in their respective quadrants about 20 degrees.

The azimuth antenna sweeps left and right while the elevation
antenna sweeps up and down. The motion translates into radar information that is similar to a gun sight. This type of antenna system
is predominantly military, with the U.S. Army, Navy, and Marines the
primary users. These types of radar are very compact, as illustrated in
Fig. 3-48. This is a AN/SPN-35 installation on board an aircraft carrier.
This is an excellent view, as it shows four types of antennae. Above
and behind the SPN- 35 is an AN/SPS-43 long-range air-search radar
antenna. This type is a broadside array and consists of wire mesh and

3-46 *Air search radar antenna showing feedhorn.* Historical Electronics Museum

dipoles. Mounted on top of the AN/SPS-43 antenna is an IFF antenna with a bar antenna in the center. Below the SPS-43 is a fire-control antenna, a true paraboloid.

A variation of the truncated paraboloid is the orange-peel antenna, which is derived from a circle. The reflector is narrow in the horizontal plane and wide in the vertical, which produces a beam pattern wide in the horizontal and narrow in the vertical. The resulting beam shape resembles a beaver tail. This type was once associated with early height-finding radar. Because more efficient antennae have been developed, such as phased arrays, this antenna is rarely found today.

The cylindrical paraboloid has a parabolic cross section in one dimension. The reflector is directive in only one plane. RF is fed by either a waveguide slit, waveguide radiator, or linear array of dipoles. This type of reflector has a series of focal points, forming a straight line. RF radiators produce a directed beam of energy. Air traffic control radar such as the AN/FPN-63 precision approach radar use this

3-47 *AN/SPN-35 antenna system.* ITT Gilfillan, a unit of ITT Defense and Electronics

3-48 *AN/SPN-35 and AN/SPS-43 antenna systems.* ITT Gilfillan, a unit of ITT Defense and Electronics

type of antenna. Different beam patterns can be developed by physically changing the width of the parabolic section.

Figure 3-49 depicts the type of shelter that a radar such as the AN/FPN-63 would be installed in. The unit is placed alongside the runway to give an unobstructed view of touchdown. The function of the system is to provide accurate landing instructions to aircraft under all weather conditions. The antenna mounted on its side provides azimuth information to the approach controller. The antenna sweeps from left to right and can be moved in the vertical plane from 0 degrees to about 12 degrees. On the right side of the unit is the elevation antenna, which is aligned to provide elevation data. The antenna sweeps up and down and can be moved in the horizontal plane about 20 degrees. Using two antennae, the resulting radar beam pattern is exactly like a gun sight.

3-49 *AN/FPN-63 precision approach radar system.* ITT Gilfillan, a unit of ITT Defense and Electronics

Both antenna beams are mechanically scanned in their respective planes. Figure 3-50 illustrates how mechanical movement is translated into beam motion. Each antenna is constructed from a series of dipole radiators mounted in a waveguide with one movable wall. A drive motor causes the wall to move in and out at a precise rate. The only

Movable waveguide wall

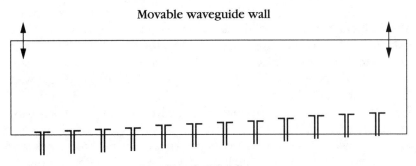

Dipole radiating elements

3-50 *AN/FPN-63 cylindrical paraboloid RF radiators.*

time a dipole will radiate or receive RF energy is when it is exactly, or very close to, one-quarter wavelength from the moving wall. As the moving wall is in constant motion, the active dipole constantly changes. That means that only one or two dipoles will be active at any given time, with the result that the beam is moved through its respective plane.

The corner reflector is very simple in construction, as it consists of two flat conducting sheets formed into a corner. The linear, or broadside array, used in VHF radar is becoming less common today. Its electrical characteristics are illustrated in Fig. 3-51. It has been replaced by parabolic reflectors and phased arrays. It consists of one-half wave dipole elements and a flat reflector. The radiating elements are one-half wavelength apart and parallel, which allows them to be additive. The flat radiator is one-eighth wavelength behind the

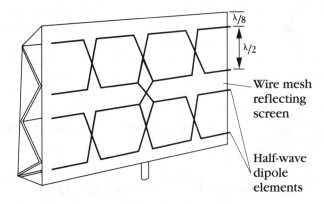

3-51 *Representative broadside array radar antenna.*

dipoles. The resulting radiation pattern is broadside to the plane of the elements. However, with the advent of stealth technology and its adaptation to aircraft, missiles, and ships, future radar systems might return to lower frequencies and the broadside array as the antenna of choice. That is because lower-frequency radar are capable of thwarting antiradar designs.

Horn radiators are capable of directive radiation at microwave frequencies. As this type of antenna doesn't use radiating elements, it has the advantage of a wide frequency band. The radiator also serves a combined impedance-matching device and directional radiator. Very flexible, it can be fed by coaxial cable or waveguide. Because it can be constructed in different shapes, as illustrated in Fig. 3-52, the resulting beam pattern is determined by the shape of the horn, its length, and mouth. The ratio of length to mouth opening determines beam angle and directivity. Essentially, the larger the opening, the more directive the radiator.

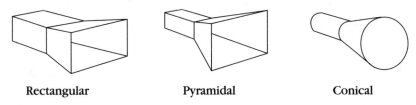

   Rectangular   Pyramidal   Conical

3-52 *Typical horn radiators.*

The phased-array antenna is gaining in popularity as computers and solid-state radiating elements continue to decrease in cost and increase in reliability. The phased array eliminates the large, complex mechanical antennae and provides "rotation" by electronically moving the beam. Figure 3-53 is a photograph of one of the more impressive radar in use today, the PAVE PAWS. Designed to detect incoming submarine-launched missiles, it is a UHF, solid-state radar system. The advantage of a phased array is that the computer system can customize the beam pattern to optimize operation. If a single beam detects a target, the computer can almost instantaneously schedule other beams to illuminate the area of interest. Because of the automatic reaction time, target parameters can often be established in seconds.

Antennae construction

Antennae can be fabricated from sheet metal, metal mesh, or carbon fiber composites. Mesh is desirable because it features a low wind re-

3-53 *PAVE PAWS phased array antenna installation.* Historical Electronics Museum

sistance. Carbon fiber is gaining in use because it is light, rigid, corrosion resistant, and it forms very accurate beams. Often, antennae subjected to extreme weather conditions are enclosed in fiberglass radomes. Radomes are useful for protecting the antenna from the elements and eliminating wind resistance as a factor in rotational speed accuracy. High winds can cause load fluctuations on the drive train and motor.

Airborne antennae are used for weapons control, search, navigation, and weather. Figure 3-54 is a photograph of an airborne radar system opened for maintenance. The antenna is housed inside a nonconducting radome for equipment protection and aerodynamics, and can be seen on the right side of the picture. Protection is essential because airborne systems are exposed to a wide range of temperatures, humidities, and pressures. Equipment assemblies and waveguides are hermetically sealed and pressurized to prevent moisture accumulation. The low air pressures encountered at altitude are conducive to high-voltage arcing. Also, the systems must be capable of withstanding severe vibration and shock.

3-54 *Airborne radar antenna.* Historical Electronics Museum

Antennae and
waveguide maintenance

Usually, the antenna and waveguides are the low-maintenance sections of the radar installation. Periodically, a rotating antenna must be inspected and have the oil changed in the gear train. Failure to maintain the proper oil level will result in a defective antenna drive system. During the inspection, the material condition of the antenna reflector, pedestal, and motor must be checked. Weather can deteriorate metal and electrical cables.

Part of the inspection is to verify antenna alignment. All rotational antennae must be referenced to a fixed point to ensure bearing accuracy. On shore-based systems, the pedestal has a true north benchmark. The antenna is stopped and manually aligned to the benchmark. At that time, the system is placed in standby. With the antenna aligned to true north, the stationary sweep on all radar indicators should be pointing north. If any are not, then the indicators need to be aligned.

Shipboard radar have to be aligned to both north and the ship's bow. The antenna has a microswitch aligned with the bow. When the antenna passes or stops on that point, a ship's head is generated. A ship's head is a mark that indicates the bow. For true north, the antenna is aligned to the ship's gyrocompass. A switching arrangement is provided so that the operator can select either true north or relative as the antenna's reference point.

Antennae and rotating choke joints can become defective. There are no locally repairable parts within either assembly. If either one

3-55 *AN/SPS-48E three-dimensional radar antenna.* ITT Gilfillan, a unit of ITT Defense and Electronics

fails to operate, it must be replaced. An antenna replacement, as illustrated by Fig. 3-55, can be a labor-intensive operation. Figure 3-55 is a photograph of the replacement of a shore-based AN/SPS-48E radar antenna. The component is so large and heavy that the services of a crane are required. Although not visible, in addition to the four people on the antenna deck, at least four others are on the ground. Because the surface of the antenna is so large, it will act like a sail in a wind. That is why both sides have guy lines attached to prevent it from oscillating.

Radomes do require some attention to ensure a long service life. To maintain the material, it must be periodically painted. Use only the type of paint recommended by the manufacturer. The wrong kind of paint can transform an electrically invisible radome into a radar reflector.

The waveguide is also a low-maintenance component. Primarily, ensure that it is painted to protect it from corrosion. If the joints are improperly sealed, moisture can enter the system. I only saw that happen once. Although water and condensation had entered the waveguide, the radar continued to function properly. The maintenance technicians noticed that system performance had begun to deteriorate. What puzzled them was that all system parameters were normal. Transmitter power out, receiver sensitivity, and system frequency were all within standards. Symptoms such as that suggested either a bad antenna or rotating joint. I was called in to verify the failure. As part of my inspection, I opened the waveguide so that I could inspect the interior. So much moisture had entered the waveguide system that the entire inner surface was severely corroded. The corrosion had attenuated the transmitted and received RF to the point that the system no longer functioned, but from outward tests appeared normal. With components such as waveguides and antennae, where failures are not readily apparent, attention to detail is important.

4

Radar receivers

The function of a radar receiver is to accept the RF signal input from the antenna system and process it so that intelligence derived from a target echo can be used by an operator. Because of the frequencies used, microwave communications receivers are very similar in concept, technology, and design to radar receivers. Early radar receivers were nothing more than modified standard communications equipment. Advances in electronics and computer processing have made modern radar receivers a catalog of leading-edge technology.

Several receiver characteristics that determine the design of a particular radar receiver include: noise, distortion, gain, tuning, and blocking. Noise is an old term that is still in use with communications receivers, related equipment, audio, and video systems. With communications and video equipment, noise is easily detected by the operator because it can be heard and seen. In modern radar operation, noise is evident by erratic video or intensities on the indicator. It can range from the almost imperceptible to completely saturating an entire quadrant.

The problem with noise is that it can mask a small target, such as a light aircraft, small boat, navigation buoy, or topographical hazard, creating a possible disaster. If noise were not a factor in radar operation, then a system's maximum range would be greatly increased to virtually infinity. However, noise, which is a factor in all receivers, limits sensitivity and range. Very small targets can often return an echo that is smaller than the internal noise generated by receiver circuits. Increased amplification, although it sounds like an alternative, is not a viable alternative because both the small echo and the receiver noise would be amplified to the same level. Reducing receiver bandwidth can also reduce noise, but it is not a desirable option either because it leads to signal losses. Radar receiver noise is predominantly thermal noise. Thermal noise results from the normal operation of resistive components and PN junctions that form the receiver circuits and components. External noise sources, other than electronics warfare by an adversary, are not very common. One exception is the interference that results when two similar radars are operated in close proximity.

The power of the reflected RF energy from small or distant targets is very low. While the returned signal strength is measured in the microvolt range, a radar indicator must have an input video signal of about 5 volts for proper operation. As a consequence, a radar receiver must have an excellent gain factor to accurately boost the video signal level to the required value. Careful design is called for, as obtaining such a high gain factor can lead to problems. To prevent undesirable obstacles such as regeneration, chassis and circuits must be shielded. *Regeneration* is positive feedback and can cause amplifier circuits to oscillate, losing echoes in large blocks of noise. Additionally, to further limit unacceptable interactions, the receiver circuits and the power supplies that feed them must be electrically isolated. Failure to do so leads to unwanted signals appearing throughout the receiver section. Electrical isolation is accomplished through the use of filter and decoupling capacitors. If one should fail, interference could appear throughout the receiver. Finding the defective capacitor could be very time-consuming.

Unlike communications receivers, tuning range is limited in a radar receiver. The small frequency variations encountered in transmitter and receiver operation call for only slight tuning changes. Even though the frequency changes are limited, they occur too rapidly to be compensated for by the operator. Automatic frequency adjustments in the form of AFC signals ensure that the system frequency remains stable. Finally, the receiver must be located as close to the antenna system as is practical to minimize any RF noise entering the system prior to amplification and processing.

Blocking is an occurrence that affects receiver sensitivity. It is caused when a received echo is so large in amplitude that amplifier stages in the receiver are overdriven. After the large pulse passes, there is a certain time interval required for the amplifier stages to return to normal operation. As a consequence, the receiver might be unable to detect smaller echoes, as they will fall below the bias level the amplifiers are temporarily held at. In this state, the receiver is said to be *blocked*. The length of time required for the amplifiers to return to normal might range from less than a microsecond to several hundred microseconds. As the detection of weak echoes is always important, a high-quality receiver must have a very short recovery time. Through design techniques, the blocking problem can be minimized.

Receiver block
diagram and theory

The most common type of radar receiver in use today is the super-heterodyne. In a superheterodyne receiver, the very high input frequency is mixed with a lower frequency to produce a third fixed frequency, called the *intermediate frequency* (IF). The IF frequency is detected, processed, and amplified to convert it to a video signal suitable for display on a radar indicator. A fixed IF frequency is used because it allows the radar designers to optimize receiver circuitry for maximum sensitivity and gain.

Figure 4-1 is a block diagram of the representative radar receiver. Returned echoes picked by the antenna system are routed to the input of the receiver via the waveguide system and duplexer. The duplexer prevents the high energy output pulse developed by the RF generator from entering the receiver and causing component damage. It also disconnects the transmitter from the antenna during receive time to prevent signal loss. The first or input stage can vary due to the age, complexity, application, and cost of the installation.

Newer designs typically use a low noise amplifier (LNA) to provide an added stage of low noise amplification to weak echoes. A parametric amplifier, tunnel diode, or microwave transistor amplifier are the usual types of LNAs found in service today. These devices are important because a decrease in receiver noise has the same effect on system efficiency as an increase in RF generator output power. Due to several factors, including cost, power consumption, crowded electronic spectrum, and equipment size, the reduced noise is more desirable.

Older systems have the RF applied directly to the first stage, which can be called either a mixer or first detector stage. The function of this circuit is to convert the high frequency of the received RF to a much lower frequency. While the high frequencies used by radar are necessary for the long-distance transmission of electromagnetic radiation, they are far too high for processing and display purposes. The lower intermediate frequency (IF) is much easier to amplify, process, and extract useful information from. The high-frequency input is converted to the much lower IF through the use of an operation called *heterodyning*. In heterodyning, two frequencies are mixed together, producing four resultant frequencies: the two original ones, the sum frequency, and the difference frequency. Typically, it is the difference frequency that is used as the IF frequency in radar and communications applications.

Heterodyning is accomplished by mixing the input RF with the local oscillator (LO) frequency. To maintain system accuracy, the con-

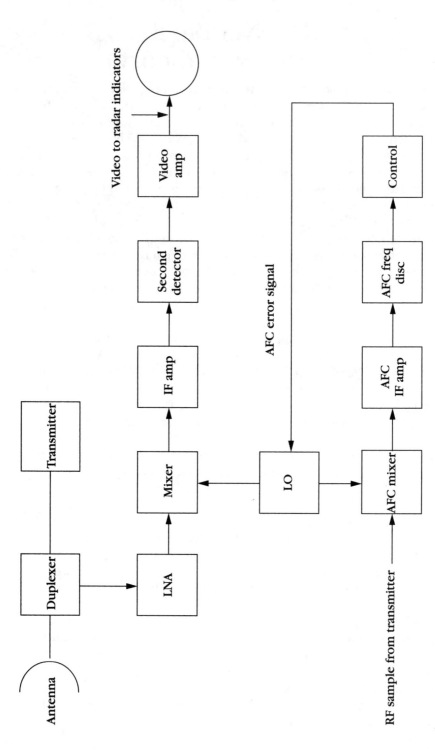

4-1 *Representative superheterodyne microwave receiver.*

version process must be concluded without signal loss or distortion. The local oscillator is a crucial receiver component, as any tuning or frequency adjustments required to lock the receiver and transmitter frequencies together are actually controlled by it. Automatic frequency control (AFC) circuits are used to keep the frequency stable within system parameters because needed adjustments often occur too rapidly for a human operator to perform for any length of time. The AFC signal is developed by comparing a sample of the transmitted RF with the LO frequency. If the transmitter frequency drifts, the AFC circuitry produces an error signal. The error signal is applied to the LO, which in turn changes the receiver frequency too, causing it to track the transmitter. By combining the input RF with the variable LO signal, the result is stable IF signal. A fixed frequency is desirable because it simplifies receiver design and allows receiver circuitry to be optimized for maximum sensitivity and gain.

The output signal of the mixer is now called the IF signal and is applied to the IF amplifier stages for amplification. The IF stages provide most of the receiver gain and determine overall receiver bandwidth. An important design consideration is that the IF circuitry must preserve the phase information contained in the returned echo for further processing and information extraction.

From the IF strip, the greatly amplified signal is applied to the second detector. The second detector actually rectifies the IF signal to derive the video information. At this point, the high frequency carrier is removed, leaving only the modulation envelope. The resulting video is then amplified by several stages of video amplifiers. The video is routed through cables to the radar displays for use by the operators.

Receiver theory of operation

The low noise amplifier is used to provide amplification while producing very little thermal noise associated with electronic circuitry. Unlike the noise of communications and TV receivers that you are familiar with, in radar applications, noise is interference that is caused by internal circuitry rather than other external electronic sources. All electronic systems do generate internal noise, but it is the external noise that is the most noticeable. Internal electronic noise is caused by solder joints and the semiconductor junctions found in transistors and diodes. The first LNAs were TWT tubes, but have since been replaced by solid-state devices that feature very low noise figures and excellent gain characteristics. As effective as TWT are, the solid-state replacements have the additional advantage of being very reliable and easy to maintain, and they require less support circuitry.

The local oscillator is a very important tube. A typical superheterodyne radar receiver has an intermediate frequency of either 60 or 30 MHz. The received RF is much higher, possibly in the GHz range. The much lower frequency signal provided by the local oscillator heterodynes, or mixes with the received RF to produce the desired intermediate frequency, or IF. The mixer actually produces four different frequencies: the receiver RF, the LO frequency, the sum of the two, and the difference of the two. Bandpass filters on the output of the mixer stage pass only the desired difference frequency, shunting the others to ground. The LO must be very stable and tunable over a small range to compensate for changes in transmitter frequency. The mixer output will be small due to the very low signal levels present in the mixer stage.

A very common LO is the reflex klystron, illustrated in Fig. 4-2. Rugged and stable, it is automatically tunable over a small range by varying its control voltage, which is known as the *repeller voltage*. A high degree of stability is vital because if an LO operated at a fre-

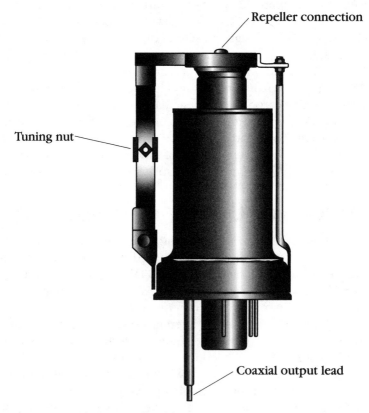

4-2 *Internal cavity reflex klystron.*

quency of 3000 MHz, a drift of only 0.1% would be 3 MHz, which is the bandwidth of most radar receivers. That small a drift would cause a significant decrease in radar system operation and loss of targets. Course frequency tuning is accomplished by varying the physical size of the klystron's cavity, as the tube is a resonant device. Cavity size is changed through the use of a movable plate controlled by a tuning nut.

A word of caution: it is advisable to adjust the tuning nut with an insulated tool to prevent any electrical shock hazard. Any automatic adjustments by the AFC circuit to link the receiver frequency to that of the transmitter is achieved by changing the repeller voltage. In summary, cavity size is course tuning, and repeller voltage is fine tuning.

An AFC signal is used to keep the LO locked 30 or 60 MHz from the received RF. Figure 4-3 is a block diagram of a typical AFC function. The basis of the AFC signal is the RF generator's output. A sample of the transmitted RF is applied to the AFC function. Therefore, the AFC mixer heterodynes a sample of the transmitter's RF output with the LO signal. The resultant signal is amplified and applied to the discriminator. A *discriminator* is a circuit that converts a frequency- or phase-modulated carrier into a signal that is amplitude modulated. Actually, in a radar system, a discriminator functions as an error signal generator. It has an output voltage that is proportional in polarity and amplitude to any shift in IF frequency. If the output of the AFC mixer is the exact value of the IF frequency, then the circuit will not produce an output. If the mixer produces a difference frequency that is greater than the IF, that results in the generation of positive output pulses. Conversely, a difference frequency lower than the IF would develop negative pulses. If an error signal is developed, it is amplified and ap-

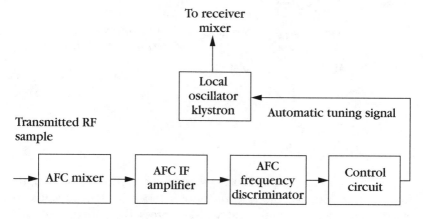

4-3 *AFC block diagram.*

plied to the reflex klystron's repeller as a control voltage. The application of an error signal changes the klystron frequency, causing the receiver to track the transmitter. Through circuitry such as this, the frequencies of the transmitter and receiver are locked together to ensure accurate system operation.

Many older radar systems use a crystal mixer circuit rather than an LNA at the receiver input stage. Vacuum tubes were not considered suitable components, as the design would induce an excessive noise level into the radar system. The input stages are an especially harmful point to have any interference or noise induced into the system, as the noise would be amplified by the following stages, possibly masking weaker echoes. Another negative point is that at microwave frequencies, vacuum tubes have an excessive *transit time*, the time it takes electrons to travel from cathode to plate.

A very common crystal is the point contact silicon crystal diode, illustrated in Figs. 4-4 and 4-5. Figure 4-4 is the internal construction of the device. For operation, it depends upon the pressure resulting from the contact between a semiconductor crystal and electrical point. As shown in the figure, the device consists of two metal posts connected by a cat whisker. Mounted on one of the posts is the semiconductor crystal.

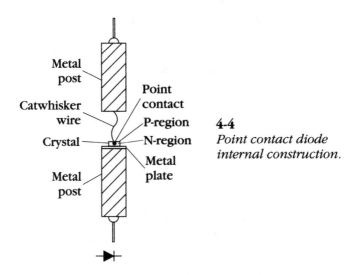

4-4
Point contact diode internal construction.

Figure 4-5 is an expanded drawing of the crystal. As shown, the largest section of the diode consists of a block on N-type semiconductor material. Notice the cat whisker pressed against the crystal. The cat whisker is a thin wire fabricated from tungsten, beryllium-

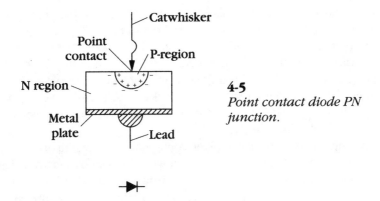

4-5
Point contact diode PN junction.

copper, or bronze-phosphor. During the manufacturing process, a large current is passed through the cat whisker to the N-type semiconductor material. The small region of P-type material is formed at that time. The result is a PN junction diode.

The point contact diode is used in microwave receivers for a very important reason. All junction diodes have a certain value of inherent capacitance due to the electrical nature of the PN junction formed between the blocks of P and N material. The electrical characteristics of a PN junction diode are depicted in Fig. 4-6. The depletion region that separates the two blocks of semiconductor material functions as the dielectric of a capacitor. As a reminder, the depletion region is the part of a semiconductor device that is lacking in current carriers, both holes and electrons. The points where the depletion region joins the P and N material correspond to the capacitor plates. The cat-whisker-junction construction of the point contact diode results in a smaller natural capacitance than a conventional PN junction diode. A smaller capacitance translates into a smaller capacitive reactance at microwave frequencies, which in turn results in a smaller value of capacitive reactive

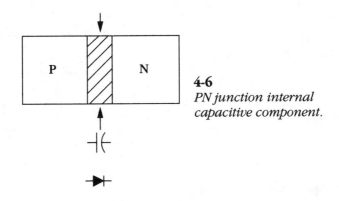

4-6
PN junction internal capacitive component.

current flow. In a forward bias state, the resistance of the point contact diode is greater than that of a conventional diode. Reverse bias current flow is more dependent upon applied voltage than a normal diode. Another feature of the point contact diode is that the device can be taken apart, as one end is a flanged mount to hold in place. It can be taken apart and put back together backwards. If you have a diode that checks good resistively but does not work in the mixer, make sure that it is assembled correctly.

The most common diode mixer configuration is an unbalanced or single-ended crystal mixer. A representative schematic is drawn in Fig. 4-7. The configuration uses a tuned transmission line that is one-half wavelength long. The function of the transmission line is to impedance match the mixer section to the LO and echo inputs. As shown in Fig. 4-7, the received energy is applied to a coil that is inductively coupled to a second coil. The second coil provides the LO input. The two signals are mixed by the crystal. The actual assembly is depicted in Fig. 4-8. Typically, a probe is used to inject the LO input to the mixer stage. The mixer section and waveguide are connected by means of a slot. Any undesired signals resulting from the mixing process, such as the carrier, LO, and sum frequencies, are eliminated by a resonant circuit tuned to the IF frequency. Although it has the major advantage of simplicity, it is outweighed by the disadvantage of an inability to cancel any LO noise. By nature, LO klystrons are very noisy, and failure to eliminate them can result in the loss of weak echoes from distant or small targets.

An improvement on the single-ended mixer is called the *hybrid*, or balanced mixer, which is also known by a more colorful name, "magic T." The schematic representation of a magic T is shown in Fig. 4-9. The term magic T came about because the assembly is shaped

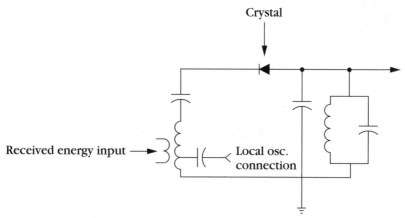

4-7 *Single-ended crystal mixer.*

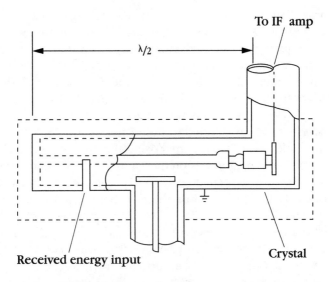

4-8 *Single-ended crystal mixer construction.*

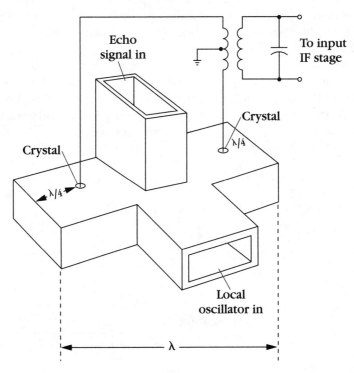

4-9 *Balanced hybrid crystal mixer.*

like a large T and, according to some people, works like magic. In this arrangement, two crystals are mounted directly in the waveguide, located one-quarter wavelength from ends of the T. The one-quarter wavelength mounting point was selected because it is the electrical point of maximum voltage. The ends of the T are electrical short circuits. The outputs of the crystals are connected to a balanced transformer to provide a path to extract the signal. For further matching, the transformer output is tuned to the IF frequency.

"T" operation will be explained using the next three drawings. In Fig. 4-10, the LO signal input is on the T base and splits into two equal paths, one for each crystal diode. The LO signal is in phase across both diodes. That is indicated by the fact that the arrows all point in the same direction. Look at Fig. 4-11. Received RF energy is applied to the echo arm. Due to the construction and electrical characteristics of the magic T, returned RF is in phase across one crystal, and out of phase across the other, as indicated by the arrows. The LO energy is always in phase across both crystals. As the received RF is in phase across one and out of phase across the other, at certain times, the two signals felt across one crystal will be in phase and out of phase across the other. That is illustrated in Fig. 4-12. The result is that one crystal will produce a positive polarity signal, and the other crystal will produce a negative polarity. As the outputs of both crystals are connected to the balanced transformer, when the signals are opposite in polarity, they are additive. When they are the same polarity, they cancel one another out. As any noise would be in phase across both diodes, it would cancel out, a simple but effective means of minimizing receiver noise generated by LO circuitry. For proper operation, the electrical characteristics of the crystals must be matched as closely as possible.

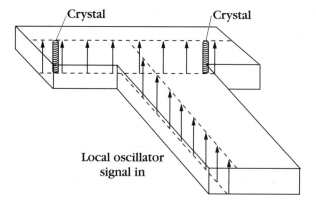

4-10 *LO input to a hybrid crystal mixer.*

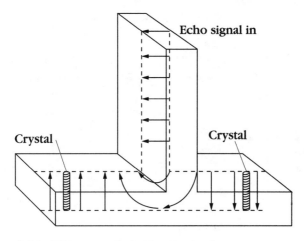

4-11 *RF input to hybrid crystal mixer.*

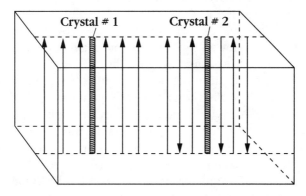

4-12 *Hybrid crystal mixer.*

Unfortunately, any noise that arrives with the received RF will not be affected, as it is in phase with it and will not be canceled out.

The IF amplifier section receives the output of the mixer section. A crucial receiver section, the IF circuitry determines system bandwidth, overall gain, and signal-to-noise ratio. The IF amplifier, or IF strip as it can be called, consists of several amplifier circuits, between three and ten stages. Generally, IF strip bandpass will be as narrow as possible without affecting the received energy, as illustrated in Fig. 4-13. Section gain must be variable to provide for the same voltage output with varied inputs from echoes of different signal strengths. As some radar systems have various pulse widths to enable the equipment to perform multiple functions, the IF strip must have sufficient bandwidth to accommodate them without loss or distortion.

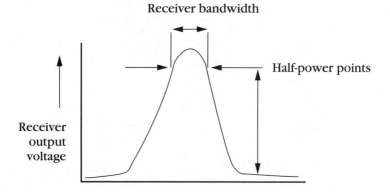

4-13 *Receiver response curve.*

No matter how many individual stages compose an IF strip, the most important one is the first. First-stage quality determines noise figure for the entire section. Any noise generated by this stage will be amplified to higher levels by subsequent stages, degrading overall system performance. As was true with the LO and input stage, any noise generated by the IF function can reduce the ability of the system to detect small or distant targets. The greatly amplified IF signal must be converted to a video signal suitable for display on a radar indicator. Figure 4-14 illustrates a typical vacuum tube IF strip. Whether an IF strip is constructed from vacuum tubes or transistors, it must be properly tuned. Notice the variable inductors connected to the input grid of each tube. IF frequency is determined by the variable inductors and the shunt capacitors. The variable inductors and the shunt capacitors form resonant circuits. Alignment is accomplished by adjusting the inductors to the point where the resonance between them and the capacitors is the proper IF frequency.

The IF function controls the bandwidth of the entire radar receiver. It must have a sufficiently wide bandwidth to encompass all the frequencies that compose an echo pulse. A narrow bandwidth causes a problem called *transient distortion*, as represented in Fig. 4-15. An insufficient bandwidth can cause a nonlinear amplification of the IF signal. The normal input is on the left side of the drawing. The nonlinear amplification causes the output waveform on the right. Notice that the rise and fall times of the waveform are sloped rather than sharp. For range accuracy, the pulses must have sharp rise and fall times. The slopes will result in inaccurate ranges, called *range ambiguities.*

As radar receivers must have a high gain factor, IF amplifier stages are normally connected in cascade configuration. Typically, a cascade

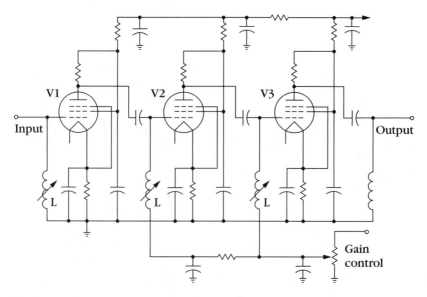

4-14 *Representative vacuum tube IF strip.*

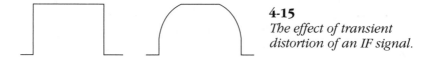

4-15
The effect of transient distortion of an IF signal.

amplifier will have a gain factor that is the product of all stages. In other words, if the function has five stages, each with a gain factor of four, total function gain would be a theoretical 1024. In addition to inducing a large value of noise, the circuit would have a reduced bandwidth, resulting in increased distortion. To counteract the bandwidth reduction, each individual stage must have an increased bandwidth. That can be accomplished in one of several ways. Figure 4-16 depicts a stagger-tuned IF strip. View A, B, and C are the bandwidth curves for a three-stage IF strip. View A is the input, or first stage. It is adjusted so that its center frequency is below the desired IF frequency. View B shows that the second IF stage is tuned to a frequency above the IF frequency. The third stage is tuned to the IF frequency. The result is the overall bandpass of the IF strip, as illustrated in the bottom waveform. It features sharp rise and fall times, adequate bandwidth, and a flat response curve, all desirable features.

Another tuning method used to increase IF strip bandwidth is double-tuned, stagger-tuned coupling. *Coupling* is when two circuits have a common impedance through which current flow in one circuit affects current flow in the second. In terms of an IF strip, the function

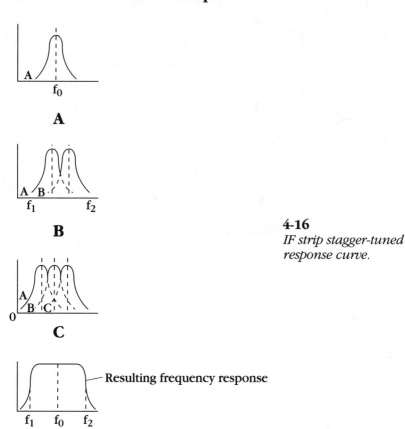

A

B

C

Resulting frequency response

D

4-16
*IF strip stagger-tuned
response curve.*

of coupling is to efficiently transfer power from one circuit to another. By double tuning, the IF strip has a flatter response across the entire bandpass and steeper sides. Figure 4-17 is the response curve for a double-tuned, stagger-tuned IF strip with varying degrees of coupling. An undercoupled circuit results in a very sharp frequency response curve. This is undesirable, as it would lead to the loss of targets. As the degree of coupling is increased, the curve resembles the transitional curve. The top becomes somewhat flattened and begins to decrease at the resonant frequency. Overcoupling results in a curve with a significant decrease of gain at the resonant frequency. Overcoupling can cause an IF amplifier to break into oscillations, an undesirable effect.

The type of coupling and alignment in a particular IF strip is determined by the design. In an amplifier with all stages double-tuned, stagger-tuned, overcoupling is not used. In addition to the before-mentioned oscillation problem, the function is very difficult to properly align. In that type of circuit, transitional coupling would be the

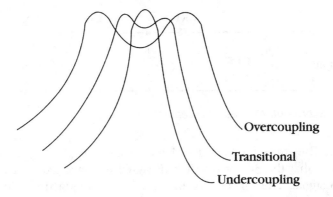

Overcoupling

Transitional

Undercoupling

4-17 *IF strip double-tuned stagger-tuned response curve.*

ideal response curve. Normally, in a multistage IF amplifier, the type of stages will be mixed. It will feature a combination of double-tuned, stagger-tuned, and single-tuned amplifier stages. The end result is a final frequency response curve similar to the stagger-tuned IF strip. The output of the IF strip is routed to the video detector circuit.

As the IF signal is unacceptable for display purposes, the function of the video detector is to transform it into a format suitable for display on a radar indicator. Figure 4-18 illustrates typical IF video. It is applied to a circuit called a video detector to extract the information. The simplest and most common radar video detector is a crystal diode, depicted in Fig. 4-19. The secondary of a transformer and a capacitor serve as a resonant circuit tuned to the IF center frequency. With no received radar signal input, the circuit does not produce an output. When a video output is passed from the IF function, positive excursions reverse-bias the diode, which results in the no-output condition. As the signal swings negative, the resulting voltage forward biases the semiconductor, allowing current to flow. The function of R1 and C2 is to convert the current flow into a voltage. C2 discharges any time the

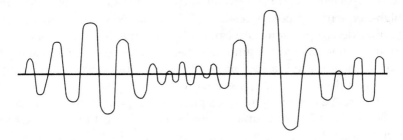

4-18 *IF video.*

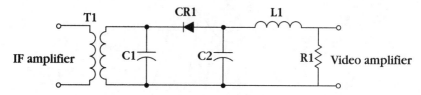

4-19 *Representative radar receiver diode detector.*

current flow through the video detector begins to swing positive. The result is that the 30- or 60-MHz IF signal is converted to a negative video signal. The function of L1, and C1, in conjunction with circuit wiring, is to form a low-pass filter to block any remaining IF components in the composite video signal. In effect, the video detector eliminates the very high frequency carrier oscillations, leaving only the modulation envelope as radar video, as illustrated in Fig. 4-20.

4-20 *Detected radar video.*

Video amplifier circuits receive the radar video from the detector circuit. Its function is to amplify the video to a level sufficient to drive the radar displays. The amplifiers are basically RC-coupled low-pass amplifiers that use high-gain active devices. Older systems would use high-gain pentode vacuum-tube amplifiers as the active components. Newer technology systems use high-gain transistors and integrated circuits. Regardless of the type of amplifier, it must feature a fairly wide frequency response with a bandwidth from 2 to 100 MHz.

To minimize losses, vacuum-tube and transistor amplifiers must use frequency compensation networks to combat interelectrode and stray capacitances. These unwanted capacitances can cause a loss in high-frequency response, resulting in a degraded receiver operation. Further design problems are caused by the coupling capacitors used to interconnect the video amplifier stages. Unless the correct value of capacitance is used, a loss in low-frequency response could result, causing a decrease in target strength.

Figure 4-21 is a schematic diagram of a representative video amplifier. C1 and C2 are coupling capacitors to connect the amplifier to other stages. Capacitors are used to provide a degree of isolation by passing only the desired signals and blocking any bias voltage levels.

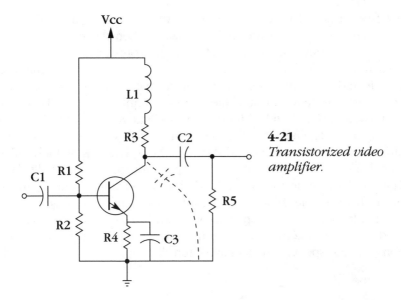

4-21
Transistorized video amplifier.

R1 and R2 set the base bias of the amplifier stage. R4 provides emitter self-bias and temperature compensations for a stable output. The function of C3 is to prevent any signals from appearing on the emitter for increased stability and gain. The function of L1 is to increase stage bandwidth to the required width without inducing distortion. The dotted capacitor represents stray capacitance formed by circuit wiring and the internal capacitance of the other components. By carefully selecting the reactive components and considering the value of stray capacitance, engineers design wideband, low-distortion, low-noise, video amplifiers. Just as the vacuum tube has been superseded by the transistor as the amplifier of choice, integrated circuits are now very common. Often, a radar wideband video amplifier will consist of a handful of ICs and resistors.

Special receiver circuits

Numerous factors, either separately or in combination, can degrade the performance of a radar system to where it is practically useless. Weather has the greatest impact on radars that operate at frequencies above 3 GHz. Weather fronts, heavy moisture, and dense cloud banks can all but render a system blind. Sea return is a problem associated with shipboard radar systems. The effect is caused by radar reflections from waves and choppy seas. During periods of severe wave action and high winds, massive blocks of radar echoes can extend out from a ship for several thousand yards, creating a large blind zone. This condition is an obvious hazard when the system is used for navigation and collision avoidance. Electromagnetic interference, both unintentional and intentional, such as an adversary jamming, can degrade system performance. Electronic clutter caused by another radar system can take the form of running rabbits. This is a problem associated with shipboard radars, rather than shore-based.

Figure 4-22 illustrates the problem. For this to occur, the two radars must be close in both distance and frequency. It will look like wheel spokes and can cover only a quadrant, or the entire display. The cure is to tune the two offending radar systems as far apart as practically possible. Blooming, caused by jamming, can trigger an entire quadrant of a radar display to be overdriven with blocks of video, rendering the system blind. To eliminate or minimize these conditions, several special circuits have been devised.

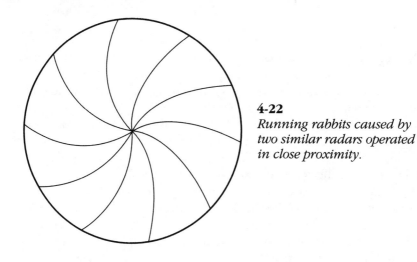

4-22
Running rabbits caused by two similar radars operated in close proximity.

Radar receivers must be capable of manipulating RF returns from different targets with widely ranging signal levels without manual adjustment. The speed of radar detection makes manual video level adjustment impossible. Manually setting the video level for close targets would most likely mean that returns from distant ones would be difficult, if not impossible, to see on the radar display. If video levels were idealized for distant returns, then close targets would cause the receiver to be overdriven or saturated. The answer is to use a special circuit called sensitivity time control, or STC. The concept is to vary receiver gain as compared to time.

In radar, distance to a target is a function of how long it takes the RF energy to make the two-way trip. STC provides an automatic exponential damping of receiver gain as compared to time. The concept is illustrated in Fig. 4-23. As shown, receiver gain is reduced at close ranges and gradually increases to normal as time in the PRT increases. By decreasing receiver gain close to the system, nearby targets are attenuated. The gradual increase to a normal gain setting allows the display of small, distant targets. The effect on a radar display is often dramatic. Figure 4-24 shows the effect of a large, close target. Notice that the echo on the indicator is elongated and very intense. An operator would have a difficult time in deciphering exactly what it was. It could be several aircraft, a heavy weather disturbance, or a receiver problem. By selecting STC, it becomes evident that only one contact is causing the echo, as depicted in Fig. 4-25. I have observed a large aircraft such as a C-5 Galaxy at close range, and the resulting echo stretched for almost an entire quadrant.

4-23 *The effects of sensitivity time control.*

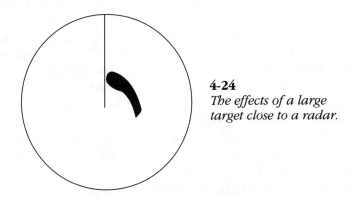

4-24
*The effects of a large
target close to a radar.*

STC circuits produce a bias voltage that controls receiver gain. Circuit action is initiated when the transmitter fires, triggering the STC circuit to reduce receiver gain to zero to prevent any leakage from the main bang being amplified. At the trailing edge of the main bang, STC circuitry begins to slowly increase the receiver gain back to normal levels. Large returns from any close targets are amplified less by the

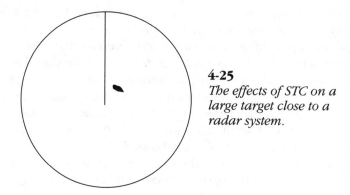

4-25
The effects of STC on a large target close to a radar system.

biased IF circuits, preventing saturation. STC usually has an effect on receiver gain for about 50 miles and is adjustable in both duration and effect on receiver gain.

Automatic gain control (AGC) is a means to control receiver gain to prevent saturation from strong targets, regardless of distance from the radar antenna. AGC normally affects the IF strip gain, and can also be known as instantaneous automatic gain control (IAGC). Different-sized aircraft will return different amounts of RF energy. For example, the returned energy from a 747 at 50 miles is almost as great as a light aircraft at 5 miles. An IAGC circuit is a wideband dc amplifier designed to allow the full amplification of weak targets. When target strength increases above the preselected threshold level, IAGC begins to attenuate the signal. The function uses degenerative feedback to provide an automatically variable negative bias and maintains signal level at the desired level of amplification. The effectiveness of AGC in a radar system is determined by the number of stages it controls. With one stage controlled, IAGC range would be limited to a factor of 20 dB. By increasing the number of IF stages controlled by the IAGC circuit, dynamic range can be increased to 40 dB. IAGC and STC are used in conjunction to ensure that all received echoes are accurately displayed without causing the receiver to saturate, obscuring the radar indicator.

Fast time constant (FTC) is another useful special feature often found in radar receivers. The circuit is nothing more than a differentiator circuit normally located at the input of the video amplifier section. When a large block of video is processed by the circuit, only the leading edge remains, and it is passed on for amplification. The circuit is designed with a short time constant so that a large block of video and a small aircraft return will produce the same-sized indication on the display. A large block of video could be caused by a massive weather front or cloud, a land mass, electronic interference, or jamming. With FTC, the effects of heavy clouds can be decreased, allowing enough of a return from an aircraft to be amplified by the receiver and displayed as useful information. Figure 4-26 shows the effects of a heavy weather form on a radar display. The result is the loss of radar surveillance in almost an entire quadrant. No matter what the system is used for, an effect such as this negates its usefulness. By selecting FTC, all returns are differentiated. As a result, only the leading edge of the cloud mass and an aircraft are displayed, as shown in Fig. 4-27. FTC does make viewing a little more difficult, but as it compensates for effects of weather and other types of interference, it is invaluable.

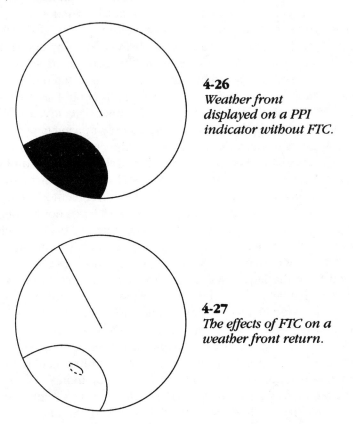

4-26
*Weather front
displayed on a PPI
indicator without FTC.*

4-27
*The effects of FTC on a
weather front return.*

The function of jamming is to reduce or completely eliminate a radar system's ability to detect an adversary's ships, aircraft, and missiles. Through its use, it is possible to attempt to saturate a radar with large blocks of video that masks assaults and intentions. Many techniques have been developed to counter jamming. Radars can operate automatically over a band of frequencies. By rapidly changing system frequency, jamming might not be able to affect it for more than two or three scans a minute. Circuits such as FTC and gated AGC can give the simplest radar some jamming resistance. Gated AGC allows large, rapidly developing signals to produce an AGC signal, providing for reduced gain.

MTI theory

As useful as a simple radar can be, clutter generated by returns from stationary targets can block areas of interest. *Clutter* is defined as extraneous echoes that can cause interference with radar operation and can be generated by vegetation, sea return, weather fronts, heavy clouds, buildings, topographical features, and other obstructions. Any aircraft overflying any of these would be invisible to the radar, as its small return would be masked by the larger return from the unwanted object. This is an intolerable situation for navigation, air traffic control, and threat detection. The solution was to develop electronic circuitry that would eliminate any stationary returns and display only moving targets.

Moving target indicator (MTI) circuitry has been in use for many years. It is special receiver circuitry that enables a system to operate in unfavorable radar environments. The MTI concept is actually very simple. The first MTI systems used the subtraction method. Stationary target removal is accomplished by a PRT-to-PRT positional comparison. Stationary targets will always have the same position, one PRT to the next. Any moving targets will have a constantly changing position, PRT to PRT. Comparison circuits invert and subtract the video received from the latest PRT with the video from the previous PRT. Any video that remains in the same position over two PRTs is canceled. If the video changes positions, indicating a moving target, then it is routed out of the cancelers to the video amplifiers.

Most modern MTI receivers, even when used with a pulse radar system, are based on the Doppler effect. As was presented in the section on modulation, the Doppler effect is most commonly experienced when listening to a train whistle. MTI uses the effect by comparing the phase of received echoes, PRT to PRT. The carrier frequency of an echo will experience a shift based on the product of the target's radial velocity and transmitted frequency. A moving target will have a changing phase relationship as compared to the transmitter frequency, from one PRT to the next. Echoes from stationary objects have the same phase relationship from one PRT to the next.

MTI block diagram and operation

Figure 4-28 is a block diagram of a representative MTI system. If you refer back to Fig. 4-1, you will notice several differences between the two. Rather than a local oscillator, an MTI radar has a STALO. The proper operation of an MTI system is dependent on a stabilized local oscillator (STALO). A STALO provides a reference of the transmitted pulse's frequency and phase. The STALO locks the receiver reference to the phase and frequency of the transmitter's output. To do this, the output of the STALO is heterodyned with a sample of the transmitter's output. The resulting signal is used to provide a lock pulse that synchronizes the phase of the coherent oscillator (COHO) with that of each transmitted pulse. The COHO provides a reference signal that is one of the inputs to a phase detector. The other phase detector input is a sample of the received video from the IF amplifiers. The resulting output of the phase detector is a video signal, the amplitude of which is a function of the phase difference between the IF video and the COHO reference. The circuit is actually comparing the phase of a transmitted pulse and the received video it generates. As polarity is not important, the resultant coherent video can be either positive or negative. The coherent video is then amplitude modulated and routed to the cancelers.

The canceler, illustrated in Fig. 4-29, consists of two separate channels: delayed and undelayed. The delayed channel includes a delay line, amplifier, and detector. The function of the delay line is to enable the received video from one PRT to be compared to that of the following PRT, so the delay line must accurately match the time interval of one PRT. This highly specialized component can be mercury or a fused quartz ultrasonic device. To ensure accuracy, many delay lines must be installed in ovens to provide a controlled environment. The undelayed channel has an attenuator that matches the losses induced by the delay line in the delayed channel. One channel will produce a positive output, the other a negative one. The bipolar video produced by the cancelers is combined into one signal, as shown in Fig. 4-30. If the target is stationary, the positive and negative videos will be equal, canceling one another out. If the videos are unequal, that indicates a phase difference in the return between two PRTs. The resulting unipolar video is routed to the amplifier circuits for display on the radar indicators.

Radar systems using a single canceler would not be very effective in eliminating all signs of clutter. Signal cancellation would possibly reduce the clutter up to two-thirds, leaving a reduced mass on the display. Many systems use double or triple cancellation to achieve the

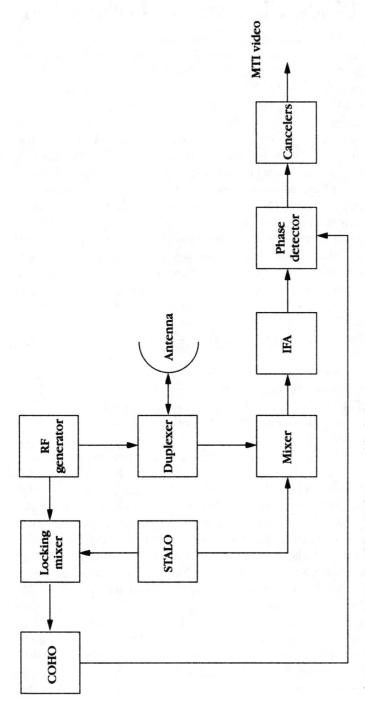

4-28 *Representative radar MTI function block diagram.*

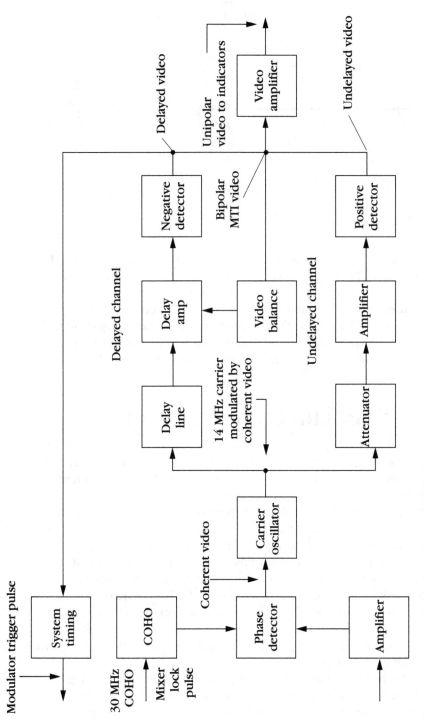

4-29 *Representative radar MTI system block diagram.*

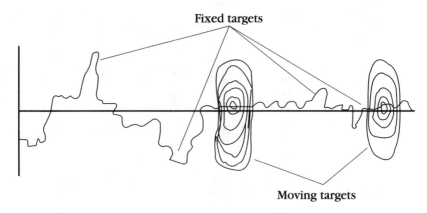

4-30 *MTI bipolar video.*

elimination of clutter from stationary targets. For example, the U.S. Navy's AN/FPN-63 precision approach radar has a double canceler arrangement that is so effective that the displays are black, as virtually all background noise and grass has been removed. For operator convenience, a circuit was added on the output of the MTI section to introduce a background sparkle on the displays. That is important, as the small amount of grass that it reintroduced tells the controllers that the system is operational.

Specialized radar receivers

A large target close to the radar can cause the receiver to saturate because of the massive block of energy it reflects. A specialized receiver has been developed that cannot be saturated. In the logarithmic receiver, a lin-log amplifier replaces the IF amplifier section of a conventional receiver. By design, this type of amplifier is a nonsaturating circuit. It accomplishes this characteristic without the use of a conventional gain control circuit. In a conventional amplifier circuit, saturation is the condition that exists when, for an increase in input voltage, there is no corresponding increase in output voltage. Saturated video causes echo displayed on a CRT to be fuzzy and indistinct. The returns appear to smear and bleed on the screen. It can also cause targets to blur together or to join with large stationary objects. Navigation would be impossible, as target edges would be indistinct, lacking any clear outline.

The response curve for a lin-log receiver is illustrated in Fig. 4-31. For small or low-amplitude input signals, the lin-log receiver has an output voltage that is the linear function of the input voltage. When a

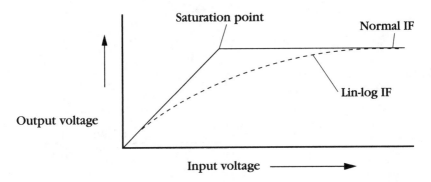

4-31 *Normal IF amplifier operation compared to a lin-log amplifier.*

large, or high-amplitude signal is processed, the output is a logarithmic function of the input. By examining the chart comparing input and output voltages, the amplifier has a seemingly exponential amplification curve. In effect, as a signal increases in amplitude, the amount that it is amplified decreases. Through the use of this technique, a very small target following a large one would often be detected, whereas in a conventional receiver, it would be lost because the saturated circuits need time to recover.

The block diagram, Fig. 4-32, illustrates one possible design for a lin-log receiver. In this example, a four-stage IF amplifier is depicted. A portion of the output of each individual IF stage is applied to a detector. The detected output signal of all stages is summed on the output of the amplifier stage. In a conventional IF strip, only the first detector would be installed. That would limit the output voltage to the saturation point of the one and only detector stage. Any further increase in echo strength would be amplified by the second stage. The resulting voltage gain would then be summed with the saturated output of the third stage. Additional target strength would be amplified by the first stage, its output being summed with the others. As a result, the amplifier never saturates, although overall gain becomes less as each stage individually saturates. Specialized IF amplifiers such as these have been in use since the early 1950s on radars such as the ITT Gifillan AN/FPN-16 precision approach radar system.

Common to automatic tracking radars is the monopulse receiver. Tracking radars are usually used for weapons control or tracking space vehicles during the launch phase. Although the system transmits a single pulse, target range, bearing, and elevation information can all be determined. A major difference between a conventional receiver and a monopulse receiver is the use of three separate channels to extract the range, bearing, and elevation data.

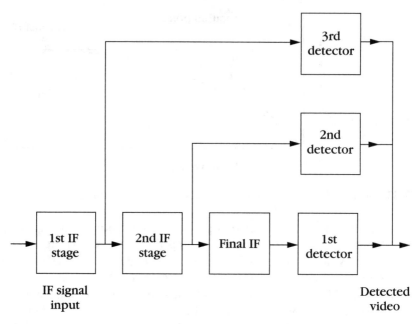

4-32 *Lin-log IF amplifier block diagram.*

To obtain the three-dimensional data, the system also uses a specialized antenna with four feed horns. The four feed horns are labeled A, B, C and D. Figure 4-33 represents the monopulse scanning technique. The position of the target within the four feed horns determines which ones return RF energy to the receiver. The system transmits energy in a circular beam pattern. The amount of energy radiated by each feed horn is equal. The amount of energy received by each feed horn depends on the location of the target in the circular beam pattern. If the target is above centerline, horns D and C receive the most reflected energy. A target that is below centerline results in horns A and B having the most energy returned. A target to the right of center returns most of the energy to horns B and D. If the target shifts to the left, then horns A and C receive the most. All positions are relative to the target, not the operator.

The RF energy received from the antenna system is routed to a summing network that provides various combinations of individual signals. Figure 4-34 is a block diagram of a representative monopulse receiver. Range information is determined conventionally, by measuring the time difference between the transmitted pulse and received echoes. Bearing data is obtained by subtracting the sum of the RF intercepted by horns B and D from the RF received by horns A and C. Bearing information is extracted by subtracting the sum of C and D

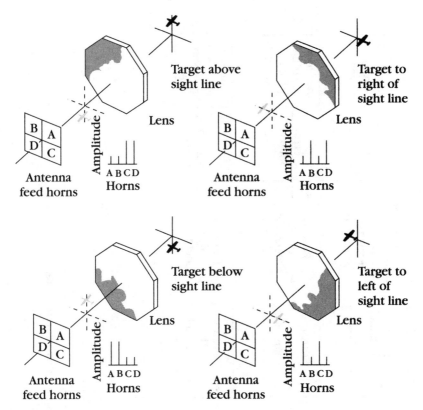

4-33 *Monopulse antenna scan pattern.*

from the sum of A and B. When the antenna is perfectly aligned in terms of target bearing and elevation, then the summing network will have a zero output. If either the antenna bearing or elevation is not aligned with the target, then an error signal is generated that will drive the antenna in the proper direction to lock on the target.

Receiver processing

Just a few short years ago, video processing was limited to special features such as AGC, FTC, STC, and MTI. Advances in computer technology and its widespread application throughout electronics has lead to amazing strides in the quality and type of information that a radar system can provide the user. Weather radars now have color displays. The color of the return identifies the severity of the weather and the thickness of the clouds. A current air traffic control radar, the ASR-9, has this feature. All controllers have the required weather information on their own indicators without needing to reference a separate weather radar

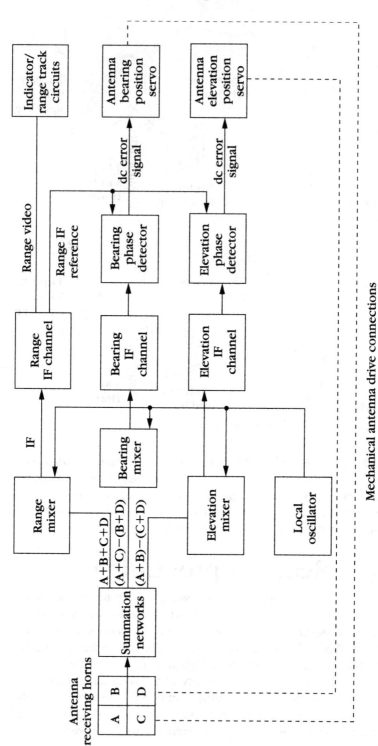

4-34 *Monopulse receiver block diagram.*

display. Other promising advances will take longer to reach widespread use due to reduced research. Unfortunately, due to the end of the cold war and a decrease in military research and development funds, advances are going to come at a slower pace.

Earlier, this book stated that a radar cannot provide a "picture" of an observed object. Targets, such as ships, missiles, and aircraft are displayed as a video spot. In 1988, a U.S. Navy cruiser shot down an Iranian civilian aircraft that was perceived as a possible threat. Surprisingly, technology has been developed that could have prevented the tragic incident. Further research could result in a radar video processor that could determine the type of target based on the pattern of the received energy.

The initial studies began in the late 1960s, as it was evident that aircraft and ships were equipped with weapons that could hit a target 50 miles away but couldn't positively identify it as hostile. Identification friend or foe (IFF) equipment was supposed to give a coded response if the target were friendly. The problem was that the IFF equipment could be inoperative or set to the wrong code. The military wanted a foolproof system to sort out the multitude of targets. The breakthrough came with the declining costs of computing power. An electronics contractor discovered that when a radar illuminated a target, every type of aircraft and missile had its own unique pattern of reflected energy. Although it is beyond current technology to display it in a meaningful way on a radar indicator, a specialized circuit could tag the return with an alphanumeric designation. In that manner, a controller would know that he was tracking a peaceful 747 airliner or a hostile attack aircraft. In today's splintered world, where naval ships often sail close to hostile nations, that type of system would provide invaluable information that could prevent future tragedies. However, due to budget cuts, research has been scaled back to almost a caretaker status.

Receiver maintenance hints

The receiver section is usually a very reliable subsystem. The most important point to remember is alignment, as that will be the most common problem. For maximum efficiency, the receiver must be properly aligned. There are no shortcuts in the alignment process. Follow the manufacturer's instructions, and you will minimize your problems. Proper alignment includes the frequency, local oscillator, FTC, STC, AGC, and MTI functions. Receiver alignment can make or break the appearance of a system. If it does not look good on the indicators, then it needs adjustment.

5

Radar displays

A radar system can only be as effective as the indicator that is used to present the information the system has obtained and processed. Radar ascertains a target's position in relation to the antenna. Subsequent echoes are used to calculate a target's velocity, range, bearing, and altitude. All of this information is useless unless it can be displayed in an intelligent and meaningful way so that an operator can interpret and use it. The encoded electronic information obtained by a radar representing target parameters is usually displayed on an indicator using a cathode-ray tube (CRT), which is similar to the television sets you are familiar with. The equipment presents the information in a visual fashion that can be easily discerned by operators and recorded for future reference.

Early displays were crude and difficult to use, which resulted in many lost targets and consequently lost opportunities in combat. The Battle of Savo Island in 1942 was such an opportunity. Four radar-equipped destroyers were stationed to provide early warning of any Japanese surface attacks. During darkness, an entire Japanese surface group slipped past the picket ships, well within radar range. Unobserved until firing, this was a costly battle to the American ships. Four cruisers were sunk, with heavy loss of life. From the ashes of defeat, radar research was pushed until today's ships are considered electronic ships. Radar display technology has progressed to the point where computer-generated symbols and color assignments to various types of targets provide the operator with a wealth of information.

Radar display classifications

The earliest radar display is called the type A scan. As shown in Fig. 5-1, there is one line or trace across the center of the CRT. This type of display is identical to what one would observe if using an oscilloscope as a radar indicator. The large block of video visible on the left side of the display is the transmitter's output pulse, or main bang. The

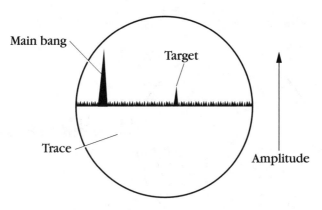

5-1 *Type A radar scan.*

sweep runs from left to right and was originally called a *trace*. The small amount of video visible along it is the receiver noise, or *grass*. Any target within the radar's beam is represented by a blip.

As with any radar target or blip, size is determined by the amount of RF energy the object reflects back to the receiver and its distance from the antenna. Unless the target is within the antenna beam pattern, the operator would never see it. Target distance is the only parameter that can be accurately ascertained, as a digital readout across the bottom of the display gives range. The amplitude of any returned echo is an indication of relative strength. A second readout gives target-bearing information, as the antenna is capable of 360-degree rotation. An A scan is difficult to visualize from two-dimensional pictures only because antenna motion drives home its inherent limitations. As the antenna rotates and the beam passes through targets, the echoes move back and forth, appear, disappear, and reappear in a most confusing manner. To the uninitiated, the display is a bewildering mass of moving blips. As can be imagined, returns from ships, aircraft, and land mass could confuse even seasoned operators. Due to the sometimes confusing information provided by the A scan radar displays, errors in navigation and threat detection happen, sometimes with catastrophic results.

The J scan is a variation of the A scan. Illustrated in Fig. 5-2, the J scan is a circular display. As shown, the trace is a circle, rather than the line that you are used to. The large blip at the top of the display is the transmitter's main bang. Any returned echoes are portrayed as blips on the circular trace. Target range is indicated by the distance of the returned echo from the main bang on the circular sweep. This type of display has the same limitations associated with the type A

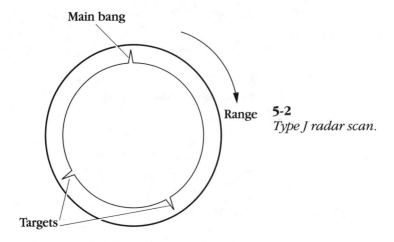

Main bang

Range   **5-2**
Type J radar scan.

Targets

scan. Not found very often in radar applications, the primary use now would be in fathometers and sonars. A *fathometer* is a small sonar that measures the depth of water under a ship's keel. In that type of application, as range to the seabed is the only target that will be encountered, its simplicity is a desirable characteristic.

The type B scan is an improvement on the A and J scans because it provides both range and azimuth information in an easy-to-use format. B scans are typically found in aviation and fire-control applications. Civilian aircraft would use this type to provide weather and collision avoidance information. Military aircraft would use it for the previous reasons, and fire control. Fire control is defined as the control and direction of weapons such as missiles, guns, and bombs.

Figure 5-3 illustrates a representative B scan found in a military aircraft. To minimize the amount of equipment that must be mounted in an aircraft, the scan has several modes of operation. The first drawing is the scan set in the search mode of operation. Azimuth information is displayed across the bottom of the display. Although up to 180 degrees can be displayed, in this instance, only 40 degrees of coverage is selected. Azimuth coverage is determined by the radar's mode of operation. Zero degrees, or the nose of the aircraft, is the zero mark in the center of the display. Range information is found in the vertical plane. The base of the display is zero range, and the farther up the sweep the echo appears, the farther away from the radar it is.

Notice that a range value is not displayed in this design, as a digital readout below provides the information. The four large lines down the center of the screen are actually the sweep, which is called *jizzle*. The position of the sweep is the actual azimuth position of the antenna relative to the nose of the aircraft. The four sweep lines are

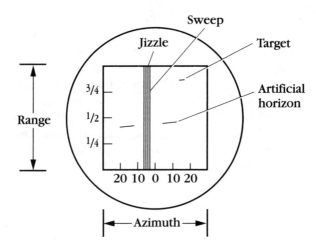

5-3 *Airborne type B radar indicator.*

generated by the movement or nutation of the antenna feed horn or reflector dish. An artificial horizon is displayed so that the pilot is aware of the physical horizon and the aircraft's position relative to it. Echoes reflected from aircraft are displayed as bright video spots on the indicator.

The display changes slightly when the radar is locked onto a target. The range information is replaced by elevation data. Figure 5-4 depicts a fire-control B scan in the target acquisition mode. The elevation data is referenced to the nose of the aircraft, which contains the radar antenna. As with the search mode, an artificial horizon ref-

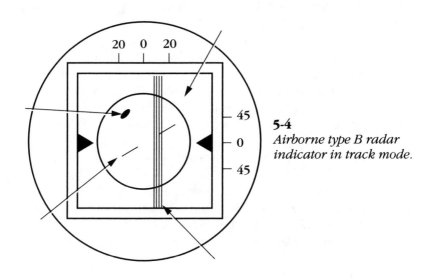

5-4
Airborne type B radar indicator in track mode.

erence line is provided. Relative range to the target being tracked is provided by the range circle, the diameter of which is proportional to the distance. In this mode, a digital readout is provided that gives the range in yards.

A type G scan, illustrated in Fig. 5-5, provides the operator with range, azimuth, and elevation error data relative to the radar. A very simple radar display, it is suitable for tracking only one target at a time. The vertical line is elevation error data, and the horizontal line represents azimuth error information. The tracked target is displayed on a line called a "wingspot." The length of the wingspot is inversely proportional to the target range.

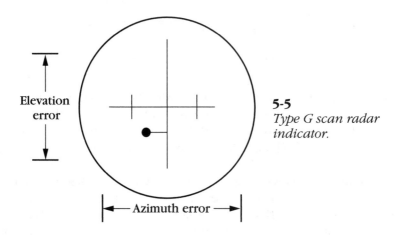

5-5
Type G scan radar indicator.

The type E scan is very useful because it displays both range and elevation data. The type E scan, as illustrated in Fig. 5-6, is very useful because it displays both range and elevation data in a format that is easily used by operators. This type of indicator is found in any application where the target's elevation is an important parameter. Range is read from left to right and altitude from bottom to top. Vertical range marks are provided to give the operator an indication of target distance. A slight variation is the EPI, or exponential position indicator, used in air traffic control (ATC) applications. Such an indicator is depicted in Fig. 5-7.

A precision approach radar is short ranged, usually 10 or 20 miles. Figure 5-7 shows that the indicator sweeps resemble the idealized charge rate graph of a capacitor. The left side of the CRT is the end of the runway or carrier flight deck. A landing aircraft would be picked up on the far right and be under direct observation and control while landing. The vertical lines on the display are the range

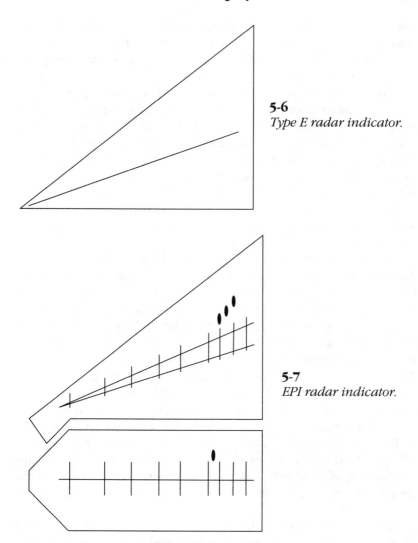

5-6
Type E radar indicator.

5-7
EPI radar indicator.

marks that the operator uses to calculate the distance that the aircraft is from touchdown, or landing. By observing them, you will notice that they are not equally spaced across the face of the CRT. From 0 to 1 mile on the left side of the indicator is much larger than the distance between the 9- and 10-mile range marks on the right. The reason is to provide a much greater degree of accuracy when the aircraft is close to landing. Due to mechanical problems and the close proximity of obstacles, landings are more hazardous than the actual flight. The greater spacing on the left gives the controller more time to provide the pilot with important course and altitude corrections.

The right-hand side of the display, which is farther away from the landing point, does not need to be as accurate because the aircraft commander has time and altitude to make any necessary corrections to line up his approach to the runway.

The display also features two separate views, as the radar has two separate antennae, one for azimuth and one for elevation information. The lower view on the display is the azimuth or bearing information. The range marks on the azimuth view indicate the position of the elevation antenna. As the elevation antenna is manually moved up and down to pick up a landing aircraft, the position of the range marks moves up and down. In the elevation view, the range marks indicate the position of the azimuth antenna. In actual use, a controller would manually move the elevation and azimuth antennae to provide the largest echo from a landing aircraft.

Notice that the azimuth scan on the EPI has a single line called the *centerline cursor*. Its function is to mark the runway centerline extended out beyond the end of the runway. It is used to give the controller an accurate indication of the correct azimuth for an aircraft to come in directly on the center of the runway. The elevation scan is different in that it has two cursors. The bottom one is called *wave off*. If an aircraft is below that line, it is too low to execute a safe landing. The upper cursor is the idealized glide slope. By keeping a landing aircraft on both the glide slope and azimuth cursors, a controller is able to give instructions for a perfect landing, regardless of visibility.

As previously stated, echo strength is dependent on the amount of RF reflected by an aircraft. Figure 5-7 is a representation of the echo returned by a landing C-5. As shown, three separate echoes are visible. One (the largest) is the energy returned by the body of the aircraft. The smaller, upper return is the tail, and the lower return is from the wheels.

The type M scan is another variation of the A scan. In this type of display, a step or notch on the trace is aligned with the target. Figures 5-8 and 5-9 depict the two types of M scan displays. In Fig. 5-8, range is determined by placing the target of interest on the edge of the range step. Distance to the target is then read off of a digital display. Figure 5-9 is the range notch variation. To obtain range, the notch is moved until the target is inside of it. Once again, a digital readout provides an operator readout. Although somewhat of an improvement, both variations suffer from the same limitations as the A scan.

The most common radar display found in shipboard and land installations is the planned position indicator, or the PPI. A typical PPI display is illustrated in Fig. 5-10. The center of the screen, called the

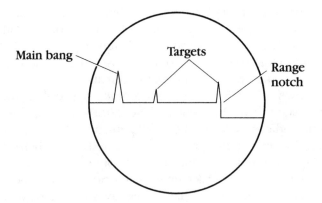

5-8 *Type M radar indicator with range step.*

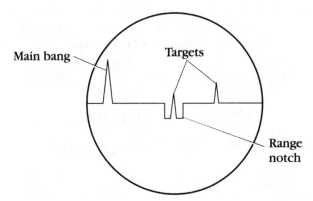

5-9 *Type M radar indicator with range notch.*

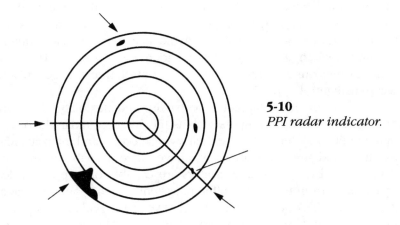

5-10
PPI radar indicator.

origin, represents the radar antenna. A line, which can be called either a trace or sweep, rotates around the center of the cathode-ray tube (CRT). The sweep indicates the antenna position. To facilitate determining target bearing, a compass rose around the edge of the CRT provides the operator with that information.

As shown in Fig. 5-10, other concentric rings, called *range rings*, indicate distance from the center of the PPI. Range rings are operator selectable and can be set on 1-, 5-, 10-, 20-, and 50-mile position. For example, if 1-mile range marks are selected, then there will be a range ring every mile. An operator-movable sweep, called a *cursor*, can be rotated around the screen so that it can be positioned over the target. A digital readout is available that gives a numeric readout of a selected target's bearing from the radar. A movable dot, called a *range bug* or *strobe*, can be moved up and down the cursor. As with the cursor, a numerical readout gives the exact range to a target. By placing the cursor and strobe over a target, accurate bearing and range information can be rapidly read off the counters on the indicator. Targets appear as bright video blips, the size of which is determined by the amount of energy reflected by the target.

Planned position indicator controls

A PPI indicator can range from a small 5-inch cathode-ray tube (CRT) or screen to a large 50-inch horizontal version. As it is the most versatile of all radar displays, it can be found in many configurations. In terms of types and numbers of radar indicators, a typical installation does not exist. A small ship might have only one, whereas an aircraft carrier would have dozens. Airports and major air route facilities would also be lavishly equipped. The simplest radar displays are used for navigation and collision avoidance. Operator adjustments are limited to range selection, cursor, and intensity controls. The more complex versions are found in air traffic control, threat detection, and weapons control applications.

Because of the equipment complexity and flexibility, there are more operator controls. Inputs such as IFF, weather, radar video, and multiple radar systems all have their own selection and intensity. Major shore and ship installations will have video available from more than one radar; therefore, indicators are capable of receiving video and triggers from up to five different systems. Even if only one radar is available, the operator can select between raw video, MTI video, map video, and secondary video. *Raw video* is unprocessed video from the receiver. *MTI video* has had the echoes from stationary objects removed, leaving only moving targets to be displayed. Based on local conditions, an operator can select the indicator to show all raw video, all MTI video, or a combination of the two. Due to a radar's beam shape, most stationary targets are close to the radar site. MTI video would only be needed for the first few miles of the sweep. Raw video would be better suited at longer ranges because any stationary objects visible to the radar beam would be a hazard to low-flying aircraft. Secondary radar is another term for IFF video. Figure 5-10 shows what IFF video would look like on a radar indicator. A video map is used to provide an easy-to-use reference on the indicator for the operator. Both IFF and mappers will be discussed in more depth later.

Figure 5-11 is a photograph of a radar indicator. This particular one is found in AN/FPN-36 radar installations. Although the FPN-36 has been replaced by newer systems, the indicator appears much the same. This type of system combines the functions of search, precision, and height-finding radar in one small package. The compass rose, which provides bearing information, is the numerical information surrounding the CRT. Operator controls include STC, FTC, antenna, high voltage, and transmitter control.

5-11
AN/FPN-36 radar
indicator. ITT Gilfillan, a unit of ITT
Defense and Electronics

STC and FTC, discussed in the receiver section, are selected to compensate for atmospheric conditions. STC helps to compensate for large targets close to the radar overdriving the receiver. As previously stated, STC reduces receiver gain at zero range and gradually returns it to normal to compensate for large echoes. FTC is used to overcome the degradation of received video caused by severe weather. Without FTC, large weather formations such as clouds could obscure a variety of targets such as aircraft and navigational hazards. Antenna controls are located across the bottom of the equipment. The switch on the left places the antenna system in one of three modes of operation: surveillance, final approach precession, and height finder.

A four-position light indicator in the center bottom informs the controller which runway the radar is observing. Just above the CRT are two bearing instruments. The one on the left indicates the antenna system position in the search mode of operation. The one on the right indicates the position of the azimuth or horizontal antenna. Other controls are for the transmitter, such as high-voltage select and transmitter on. In terms of operator controls, this is actually a relatively simple one. Because of the numerous operator controls and adjustments, it is easy to see why it takes time to become proficient in its operation.

Radar indicators designed to function with long-range systems have the ability to offset the sweep. Under normal conditions, as illustrated in Fig. 5-12, the radar antenna is referenced to the exact center of the CRT. If the CRT center were offset, it would have the desirable effect of magnifying a portion of the sweep. Notice that only one target, in the northwest quadrant, is visible. Figure 5-13 compares a radar picture under offset operation. By changing to offset and shifting the sweep origin to southwest, the target now appears at the center of the CRT. It has been magnified to the point that it now appears as two targets rather than one. Also notice that the range marks are no longer circular. That is caused by the indicator only displaying a portion of the full 360 degrees of rotation. A disadvantage would be the sweep rotation. As sweep rotation is controlled by antenna movement, one revolution around the CRT seems to take longer than when in normal operation. If you time it, you will find it to be an illusion. The only result is that the video would appear to be a little dimmer. Civilian air traffic control radars would also use this feature. Often, due to traffic congestion and navigational hazards, it is desirable to magnify a portion of the sweep.

Offset would be used in conjunction with an airborne radar system such as the U.S. Air Force's AWACS and the U.S. Navy's Hawkeye

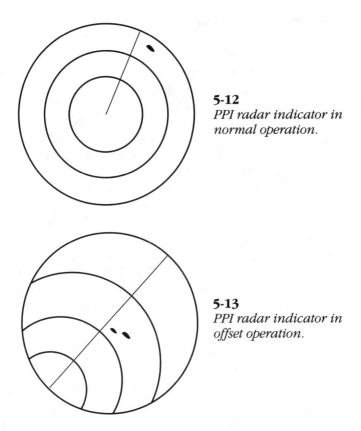

5-12
PPI radar indicator in normal operation.

5-13
PPI radar indicator in offset operation.

Airborne Early Warning (AEW) aircraft. In both examples, the aircraft flies close to an area of interest to be within radar range. The resulting video is relayed back to a ground or ship-based site for display. This has the effect of extending the radar eyes and allowing multiple sites to have access to the vital video information.

Radar indicator inputs

A radar display is not a stand-alone piece of equipment. For proper operation, it requires inputs from other functions within the radar system. First, the timing section, or trigger generator, must supply system triggers because the display must be synchronized to the rest of the radar system. When the transmitter fires, the indicator must be enabled, or turned on, so that it can display detected targets. Bearing information from the antenna system is required as another input. It is needed to align the sweep with the antenna position to provide an accurate positioning of any echoes on the display. For the bearing information to be meaningful, it must be referenced to a specific point. The bearing reference, for shore installation, is typically true north. Shipboard and aviation radars can use either true north or relative as the reference. Relative is when the antenna is referenced to the bow of the ship or the nose of an aircraft.

Video information from either the radar receiver or installed video processing units is required to input received targets. An input from identification friend or foe (IFF) equipment is required to have IFF codes available on the display for the operator's use. Many radars now use an external piece of equipment called a mapper. The function of the mapper is to overlay a video map on the radar display. Maps can be quite intricate, having symbols for runways, air routes, hazards to navigation, and many other types of information. Total system synchronization is crucial to ensure that the video is properly placed on the sweep. The PPI display is superior to all other types because it gives an accurate representation of the terrain, ships, and aircraft surrounding the system, as one would obtain from a map.

Figure 5-14 is a drawing of a radar in indicator IFF and map video. The circle in the center represents the airport where the radar is installed. The dashed lines extending from the circle are the runways. Dashed and solid lines represent restricted areas and air routes, superhighways in the sky. The target in the lower left has an old style, or raw, IFF return with it. The target on the left side of the display shows the alphanumeric type of IFF returns in use today.

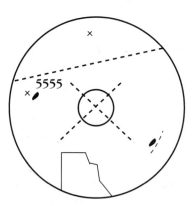

5-14
PPI indicator with IFF and map video.

Radar indicator theory of operation

When discussing radars, the best function to begin with is the gate circuitry. Gate circuitry, the block on the left side of Fig. 5-15, receives trigger inputs from the radar's trigger generator to ensure system synchronization. The type of triggers required for indicator operation is determined by the complexity of the radar system. Triggers can be as simple as just a system trigger, or could include signals such as pretrigger, system trigger, system clock, and dead time trigger. Pretrigger would preset conditions within the indicator so that all circuits begin at zero. The system trigger would initiate the sweep, and the dead time trigger would stop sweep action. The system clock would be found in digital systems and would be used to control computer circuitry that developed symbology and stored and processed digitized radar information.

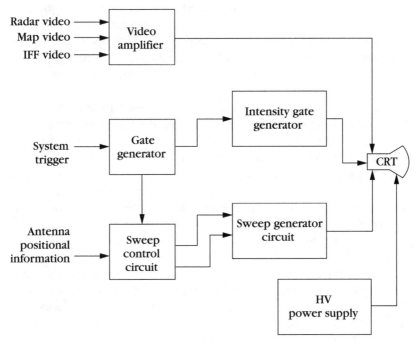

5-15 *Representative PPI indicator block diagram.*

The gate circuitry uses the trigger to develop various gates that control indicator operation. The Intensity gate serves as a bias voltage that turns on, or unblanks, the CRT during sweep time. The intensity gate is vital, as it enables the CRT to display radar information. The sweep gate enables the sweep generator, which is the circuit that develops the sweep that represents the antenna and radar beam positions.

The sweep generator, or control function, requires two inputs to operate: the sweep gate to enable it, and the mechanical position, or bearing information, from the antenna. The circuit converts the antenna information into varying voltages that correctly position the sweep on the indicator. While most radars use an electrical voltage to indicate radar antenna position, newer systems rely on digitized positional data.

The sweep generator circuit produces the voltages and currents that deflect the electron beam across the CRT to produce the sweep. To accomplish this, varying voltages and currents are applied to the deflection coil in the CRT. The gate duration and timing determine sweep rate and the maximum distance covered by each sweep. The potential developed by the circuit actually determines the sweep azimuth.

The intensity gate generator produces the gate that enables, or unblanks, the CRT during sweep periods. It also serves as a dc bias that controls sweep intensity. A CRT is a large vacuum tube. Although all signals and voltages might be present, the tube cannot operate unless it has the proper bias voltages on the control grids.

The range strobe and range marker generator are important functions to the operator. There must be a means for the operator to accurately measure distance to targets. The range strobe is the manually adjustable intensified spot that rides on the cursor. It is connected to a front panel readout that provides target distance in either yards or miles, depending on sweep range. Range marks are concentric circles appearing at fixed distances from the CRT center. If the radar is set a maximum range of 5 miles, then it would display a range mark every mile. If the range were increased to 50 miles, then it would have a range mark every 5 miles.

The final circuit is the video amplifier. It receives radar video from the receiver and other sources and amplifies it for display on the CRT. The other videos would be range marks, range strobe, cursor, and IFF, and map information would be mixed with the raw and MTI video to form a composite video signal.

A hallmark of a radar system is accurate target data. There are several ways used to ensure the accuracy of distance to targets of interest. Figure 5-16 is the block diagram of one solution, the range gate

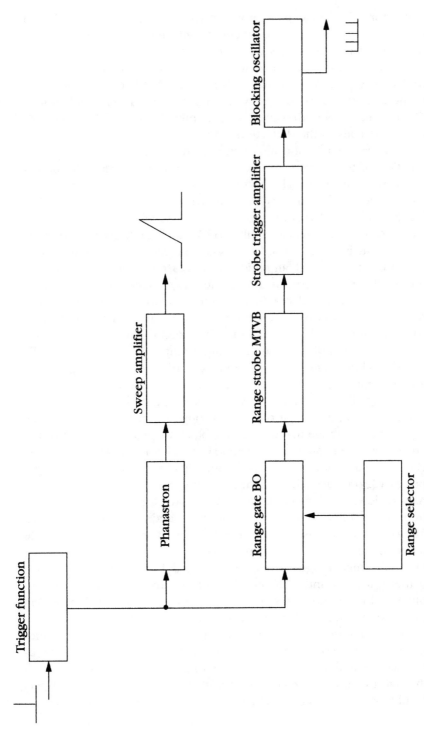

5-16 *Radar indicator range gate generator block diagram.*

generator. The range gate, not shown on the simplified indicator block, ensures an accurate sweep length and the generation of a movable marker pulse. As with most radar circuits, it begins with a system trigger for synchronization purposes. The sweep circuit is controlled by a *phantastron*, which is a circuit that is capable of developing a linear ramp voltage when triggered by an external source. The resultant sawtooth waveform is amplified and applied to the CRT deflection coils as the deflection voltage.

The trigger function also controls the range gate circuitry. The function of the range gate circuitry is to produce a variable marker that can be used to highlight the position of a target under surveillance. The trigger enables the range gate blocking oscillator. The blocking oscillator is controlled by a front panel input, the range selector. A blocking oscillator will produce a single trigger pulse when enabled. It, in turn, is used to trigger the range gate multivibrator, which is differentiated, amplified, and in turn triggers another blocking oscillator. The result is a movable range strobe that can be used to mark a target and obtain an accurate readout of distance. The range selector controls where the range strobe will appear on the cursor. A front panel digital readout is provided that gives the operator an accurate numeric indication of target range.

The other range circuit is the range mark generator. Range marks are concentric rings on the radar display that provide a means of measuring distance to a target. While the range strobe provides an accurate readout, the range rings are a rough estimate. Figure 5-17 is a representative range mark generator block diagram. The heart of the range mark generator is an 80.86-kHz oscillator. The frequency was chosen because it results in an output that consists of accurate trigger pulses 12.36 microseconds apart, which corresponds to a radar mile. As a review, a radar mile is the time it takes for RF energy to travel from the antenna to a target 1 mile away and back to the antenna. The output of the oscillator is 1-mile range marks. To obtain 5-mile range marks, a 5:1 counter counts down the 1-mile marks in a 5 to 1 ratio. Ten-mile range marks are obtained by applying the 5-mile marks to a 2:1 counter. The range marks are applied to the video amplifier. Which range marks are displayed on the indicator depends on the maximum range of the sweep.

CRT screen persistence is very important to the operator. A long-persistence phosphor coating on the inner surface of the CRT is crucial, as it will allow the radar echoes to glow after the sweep passes through them. If the screen did not glow, then the only time targets would be illuminated is when the sweep passes directly through

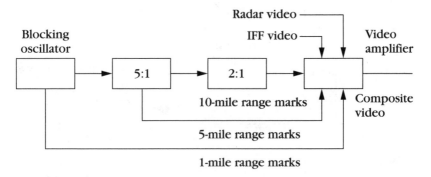

5-17 *Range mark generator block diagram.*

them, much less than a second in duration. That would be an insufficient time for an operator to observe and use the information detected by the radar.

Sweep deflection is an important aspect of radar indicator theory of operation. Early radar display designs used electrostatic deflection. Due to technological advances, electromagnetic deflection is the current method of choice. The primary difference between the two is the method of focusing and controlling the deflection of the electron beam that forms the sweep. The advantages that are associated with electromagnetic deflection include simpler cathode-ray tube construction, superior beam control, improved deflection accuracy, and better beam-positioning accuracy.

The heart of a CRT is the electron gun, which is composed of a heater, cathode, control grid, screen grid, focus coil, and anode. Figure 5-18 is a schematic representation of the internal construction of a typical CRT. The face of the CRT, or operator's viewing screen, is coated with a fluorescent material called the *aquadag*. Electron flow is from the heated cathode through the grids to the screen. The action of the electron beam striking the screen will cause a small dot to glow. A dc current flowing through the focus coils keeps the electron beam tightly bound, overcoming the electrons' repulsion. The focus coils have a potentiometer, the focus control connected in series. The focus control is an operator adjustment located on the front panel of the indicator.

For proper operation, the tube requires correct voltages, including high voltage. The high-voltage input is a connection located near the neck and the point where the CRT flares out. Located in series with the high-voltage current path is the intensity control, another front-panel adjustment. The intensity gate is normally applied through the

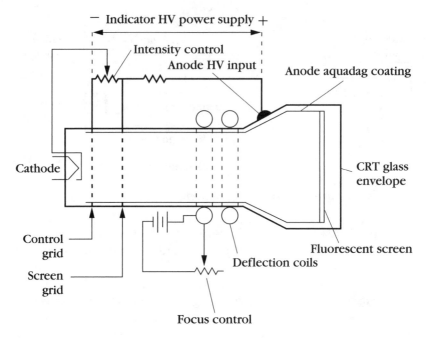

5-18 *Radar indicator cathode-ray tube internal construction and circuit connections.*

cathode circuit. Sweep and radar video are applied through the control and screen grids.

Deflection coils mounted on the neck of the CRT move the beam to form the sweep. The deflection coils are electrically an LR circuit, as depicted in Fig. 5-19. Four deflection coils connected in pairs are mounted around the neck of the CRT. The four coils are north (N), south (S), east (E), and west (W). N and S coils are connected in series, produce a magnetic field in the horizontal plane, and are called the vertical deflection coils. E and W coils are connected in series, produce a magnetic field in the vertical plane, and are the horizontal deflection coils. The name deflection coils was applied because, when under the influence of a magnetic field, an electron beam will be repelled, or deflected, at right angles. Beam deflection is proportional to the strength of the applied magnetic field. Magnetic field strength is directly proportional to the value of current flowing through the coil. As a result, CRT sweep circuitry, installed in equipment that uses electromagnetic deflection, provides deflection currents. A sawtooth current is required to develop a linear trace on the face of the CRT. Because of the natural inductance of the deflection coil, a trapezoidal voltage must be used to obtain the desired current sawtooth.

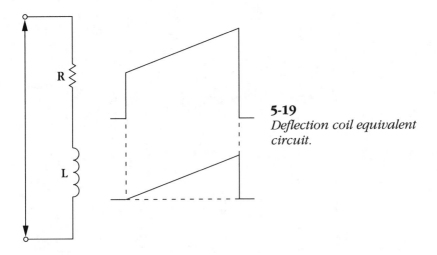

5-19
Deflection coil equivalent circuit.

Sweep rotation is required so that an indicator can display a full 360-degree view of the area surrounding a radar installation. Originally, the first PPI displays used a mechanical gear train arrangement to provide for sweep rotation. The deflection coils actually rotated around the CRT neck, moving the sweep with them. Although a very simple concept, it suffered from inherent accuracy and maintenance problems. Currently, the deflection coils are fixed, and a varying current produces an electronic sweep rotation.

Figure 5-20 depicts the deflection currents associated with rotating a sweep through a 360-degree antenna rotation. For example, at 0 degrees of antenna position, the horizontal deflection coils have 0 current flowing through them, and the vertical deflection coils have maximum current. As the antenna rotates, the deflection coil currents follow a sine wave pattern. At the 180-degree position, the horizontal deflection current is once again at 0, and the vertical deflection current is at its peak negative value.

To trace sweep rotation through the deflection coils, use Fig. 5-21. The arrow indicates the direction of the magnetic field and the trace the resulting position of the electron beam. Begin with trace positioned at 0 degrees deflection, or true north. Current flow causes a magnetic field to be developed along the axis of the coils. The CRT sweeps vertical because the electron beam is forced perpendicular by the action of the magnetic field. To begin rotation, current flow through the N-S coils decreases, and current flow begins through the E-W coils. The effect of the changing sweep currents moves the sweep to the 45-degree position. N-S current flow decreases to 0, and the horizontal or E-W current flow increases to maximum value. That action results in a sweep posi-

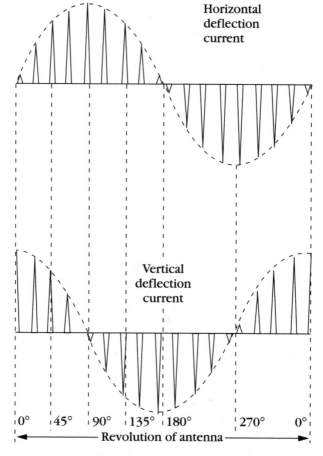

5-20 *Horizontal and vertical deflection currents compared.*

tion of 90 degrees. E-W currents begin to decrease, decreasing the magnetic field through the horizontal coils, and N-S current through the vertical coils begins to increase, causing the sweep to move to a position of 135 degrees. For accuracy, the antenna position and speed must be synchronized with the sweep motion.

Analog radar systems use synchros and servos to lock the antenna and sweep together. A *synchro* is an ac electromechanical component that is used to transmit angular position data. In construction and outward appearances, synchros and servos are very similar to ac generators. For example, the synchro mounted on the radar antenna would be known as a TX. As the antenna rotates, the position of the synchro rotor changes, producing a sine-wavelike output. The sine

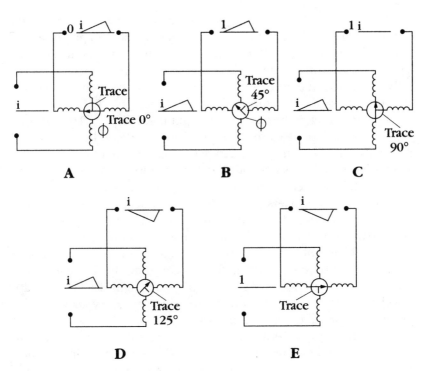

5-21 *Sweep rotation.*

wave signal indicating antenna position is routed to another synchro mounted within the radar display, called a TR. The signal from the antenna induces a changing voltage in the TR coils, which in turn develop an ac voltage that indicates the exact antenna position.

Digital-based radars use a more advanced and accurate arrangement that is based on computer circuitry. The 360-degree antenna rotation is divided up into 4096 segments represented by digital bits. As the antenna rotates, it generates an azimuth change pulse (ACP) every 0.866 degree. The reference point of true or relative north is indicated by the first pulse, which initiates rotation and is called the *azimuth reference pulse*. The digital antenna data is preferable, as it makes it easier to send antenna data over long distances.

As technology evolves, radar indicators will gain additional capabilities. Computer control promises to increase efficiency, reduce workload, and increase safety, with fewer operators. The Federal Aviation Administration is developing a highly automated radar display system that will cut the work force by up to one-half. Many repetitive tasks are handled more effectively by electronic circuitry. Sea-based indicators are benefitting from the advances as well. For several years a computer-controlled collision avoidance system has been in use on major ocean vessels. The bridge personnel set the equipment to automatically track targets. If a radar target closes to within a specified distance, an alarm is activated to alert watch standers to a possible collision. Through careful use, it can reduce the workload, while still providing a significant degree of safety.

An interesting prospect is the daylight radar display. The vast majority of radar indicators must be used in a darkened environment. The spaces are kept in virtual darkness, with only read lights and the glow from the CRTs as lighting. The first attempts at daylight displays were the BRIT systems. They were designed under an FAA initiative to provide the controllers in the tower with a means of viewing radar video. Although they controlled only visible aircraft, there were times that radar information would be beneficial. During periods of reduced visibility, the indicators were used to show the controllers an aircraft's exact position. That made it easier to positively sight it. Several European navies are beginning to use daylight displays in shipboard installations. In addition to making emergency maintenance easier, it simplifies the operators' jobs.

As technology increases, radar indicators will become more like computer monitors. That promises to make even more information easily available to operators. Current research to incorporate heads-up displays and collision avoidance systems in cars promises to make radar applications more common.

Maintenance hints

As with all equipment, radar maintenance can be broken down into corrective and preventive categories. With a radar indicator, alignment is vital. The display is the one subassembly in a radar system that everyone can observe. If a display does not look correct, it is obvious to everyone. If the cursor is placed on 0 degrees, then the digital readout should also indicate 0 degrees. If the range strobe is aligned over a range mark, then the digital readout should agree with it. Bearing accuracy should also be maintained. An error of only 2 or 3 degrees could mean a navigational catastrophe.

The alignment of precision approach radar indicators is especially crucial. The radar must be aligned to the runway to ensure accurate landings. A device called a reflector is used to mark various alignment points on the runways. As shown in Fig. 5-22, it is triangular in shape. The open end is aimed toward the antenna. Just below the reflector is a box. It contains a small oscillator. The function of the oscillator is

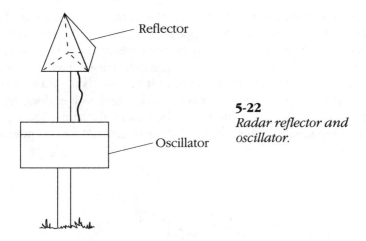

5-22
Radar reflector and oscillator.

to make the stationary reflector appear as a moving target to an MTI-equipped radar. The number of reflectors installed around a runway is variable.

Figure 5-23 is a typical installation. Notice that runway 13 has 2 reflectors, centerline and touchdown. Centerline indicates the exact center of the runway. Touchdown marks the point on the runway where the aircraft's wheels should strike. Runway 24 is slightly different. Rather than a single centerline reflector, it has two parallel to the end of the runway. The touchdown reflector is located on the opposite side of the runway. A fourth reflector is called runway parallel. It marks an imaginary line parallel to the runway. The function of these

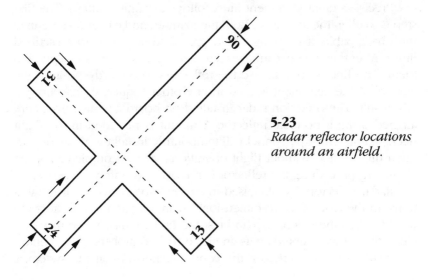

5-23
Radar reflector locations around an airfield.

reflectors is to ensure the proper alignment of the radar. Figure 5-24 illustrates how these reflectors appear on the precision approach indicator. Notice that the elevation cursor ends at the touchdown reflector. In the azimuth scan, the azimuth cursor passes directly through the end of runway reflector, as it represents the runway centerline. The touchdown reflector is off to the right, as it is alongside of the runway. No two installations will be exactly the same. The equipment is designed so that reflector placement can suit local conditions.

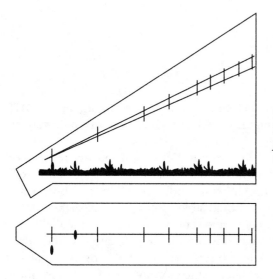

5-24
Reflectors viewed on a precision approach radar indicator.

Precision radar alignment must follow a definite pattern. The first step is to align the antennae. Both the azimuth and elevation antennae must be capable of scanning a 20-degree field of view with prescribed limits. After the antennae are correctly set, the indicators are aligned to them. The final step is to use the reflectors to correctly position the cursors. If the alignment is tried in any other sequence, inaccuracies will result. The precision radar installed on board aircraft carriers are aligned using a portable reflector. A mount is provided in the flight deck. Maintenance personnel will temporarily install the reflector during a time period without flight operations. At the completion of the alignment procedure, the reflector is removed for safety purposes.

Radar reflectors are also used in airport surveillance radar installations. In the case of a two-dimensional search radar, the reflectors are used to align the video map and test the MTI circuits. Video maps include the runway centerline as an aid for the controllers. If the centerline mark splits the reflector, the controller knows that the system is properly aligned. As the reflectors have an oscillator to mimic a moving target, it will show up, even in MTI video. The presence of the reflector video is an indication that the MTI circuits are operating normally.

Other types of radar indicators must be correctly aligned. Adjustments that must be made include range, bearing, and shape. The radar presentation must be round, not egg shaped. Sometimes, obtaining a round display can be time-consuming. The range marks must be adjusted so that a 5-mile range mark is 5 miles.

The radar indicator is useful in failure analysis throughout the entire radar system. By observing a failed radar display, you can isolate a failure to one or two functions. The first step is to look at all installed and operational radar indicators. If only one of them is inoperative, then the problem is confined with that one piece of equipment. Next, try all the front panel switches and controls. Many times, an inoperative display is nothing more than a switch in the wrong position or an intensity turned down.

Many indicator failures are easy to localize. Missing radar video, IFF video, map video, range marks, cursor, or range strobe are confined to a single function. If only one is not present, then only a few circuits can cause the fault. If all of them are missing, then the most likely cause is the video amplifier, which is the point at which all video inputs are mixed to form composite video.

A blank indicator is more complex because several different functions can cause the problem. The easiest to localize and repair would be a problem in the triggering circuits. No radar trigger means no radar indicator operation. A loss of the high-voltage power supply can give almost identical symptoms. The quickest way to separate the two possibilities is to turn the indicator off, then back on. If the high voltage is good, you will see a defocused, dim spot at the center of the CRT when power is cut off. A complication to this is that many new radar indicators have a fault detection circuit to cut off high voltage if the sweep is lost for any reason. If you have a display with this feature, then check the fault circuitry before tearing into the high-voltage section.

Triggering also includes the development of control signals and gates. The loss of an intensity gate would have a blank CRT as the most obvious symptom. If the trigger input to the range mark generator were missing, then the circuit would be unable to function.

Another common failure is sweep rotation. Either the sweep is stationary or is missing in one or two quadrants. A stationary sweep is most likely caused by a failure of the amplifier that receives the antenna rotation information. If the sweep is just missing in one or two quadrants, then the sweep drivers are the point to begin troubleshooting. A radar display has two or four sweep drivers, depending upon design. In the case of a unit with two sweep drivers, one is for north-south deflection, the other for east-west deflection. The loss of north or a north-south sweep would lead you to the north-south deflection amplifier. Although the deflection coils are a major part of the circuit, it is a rarity for them to fail. That is one problem that I have never encountered.

Planned position indicators are dependent on antenna positional information to provide the rotating sweep. Problems can arise with the information and the indicator, and this can cause sweep problems. If the system uses digital antenna information to rotate the sweep, circuit failure can make the sweep jump, stop in one quadrant, or flash. That is because the digital bits have to be processed using shift registers and counters. If a register locks up, the sweep can freeze, or jump ahead several degrees. Analog radar systems use synchros and servos to transfer the antenna information. In construction and appearance, a synchro-servo is similar to an ac generator in appearance and theory of operation. The components contain a rotor, stator, and brushes. A brush failure results in a breaking or tearing sweep. Rather than a thin distinct line, the sweep covers several degrees and resembles lightning. If you have a sweep problem, remember the circuits that process the antenna information.

Many radar indicators all seem to have a common problem, a burned mark at the center of the CRT. This is caused during the initial installation alignment or after a failure in the sweep circuits. The burned marks are caused by the undeflected stream of electrons striking the face of the CRT. To minimize or eliminate this problem, turn down all intensities when performing alignments and troubleshooting.

One problem that is all too common, but easy to correct, is accuracy. Every equipment requires preventive maintenance to ensure operation. Radar indicators need to be aligned on a regular basis. All radar applications need a correctly aligned system to meet user expectations. A range error of 5 percent is acceptable in some applications, while totally unacceptable in others. For example, a 5 percent range error would be plus or minus 50 yards at a distance of 1000 yards. Increase the system range to 100 miles, and the same 5 percent range error would be plus or minus 5 miles. The former is unacceptable in navigation and the latter is unacceptable in air traffic control.

Attention to detail is often stated as important in many professions, and radar maintenance is by no means an exception. By attention to detail I mean the small things. Panel lights are a good

5-25 *A radar indicator is how you are judged as a maintenance technician.* ITT Gilfillan, a unit of ITT Defense and Electronics

example. The bulbs are cheap and easy to replace. Yet, when you go into many radar control rooms, many burned out panel lights are evident, along with loose or missing control knobs, labels, and cleanliness. As the radar indicators are visible to all, they must be given careful maintenance and attention to detail to showcase a well-maintained radar system. Remember, as illustrated in Fig. 5-25, the appearance of a radar indicator is how you as a maintenance technician will be judged.

6

Ancillary equipment

A radar installation is not normally a stand-alone electronics system. To be integrated into shipboard or airport operations, it needs the support of additional systems. While some are to ensure equipment operation, others multiply a radar system's effectiveness and usefulness. Regardless of where a radar system is installed, it requires communications systems before it can be used. Interior communications are required to link radar operators, supervisors, and visual controllers. External communications are needed to provide communications channels among the radar operators, aircraft under control, and ships. Needless to say, without secure communications a radar system is useless.

Cooling systems

Radar requires cooling systems in order to function properly. Any electronic equipment produces a tremendous amount of waste heat, and radar is no exception. Cooling systems can be divided into two categories: ambient systems and internal heat exchangers. Many systems require both for proper operation. Equipment that is exposed to excessive heat, either internal or external, has shorter operational life spans and suffers far more component failures.

While one would associate the need for air conditioning with the transmitter section, the remainder of the system also requires cooling. Without it, a shipboard combat information center or shore-based radar room would be unbearable in a short period of time. CRTs require high voltage to operate, which translates into a great deal of heat. If left without external cooling, a radar indicator would burn itself out in a few hours. Many shipboard radar indicators are mounted in drip-proof cases to allow use in damp environments. That means the heat-producing electronic components are sealed in a metal box. To keep the equipment operational, heat exchangers are used. The idea is to maintain case integrity while still providing a means of eliminating waste heat.

Radar transmitters require air conditioning to maintain ambient room temperature and to cool internal components such as the thyratron and RF generator. At the very least, equipment cabinets will use fans to provide air flow for cooling. Without the fans, cabinet internal temperature would rapidly rise to the point where component damage would result. To prevent that from occurring, air flow sensor switches are often installed as a protection circuit. If the fan fails to operate for any reason, the switch opens, removing power from the affected cabinet.

While fans are satisfactory for most electronic equipment, RF generators such as klystrons, traveling-wave tubes, and magnetrons produce so much heat that other methods are required for satisfactory cooling. Liquid cooling is often employed as a means of reducing the temperature of major heat-producing components. Figure 6-1 is a simplified drawing of a representative cooling system that would be suitable for a radar installation. As shown, the system consists of two separate parts: the primary loop and the secondary loop. The primary loop provides the source of cooling action. The origin of chilled water for the primary loop could be a major air-conditioning system, a chill water system, or in the case of shipboard installations, the ocean itself. The two loops are isolated from one another to ensure that the purity level of the water in the secondary loop is maintained.

Figure 6-2 is an expanded drawing of the secondary loop. This type of system would be suitable for a klystron-based radar installa-

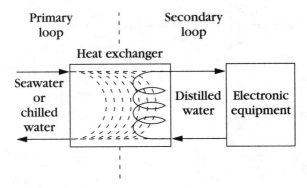

6-1 *Radar cooling system block diagram.*

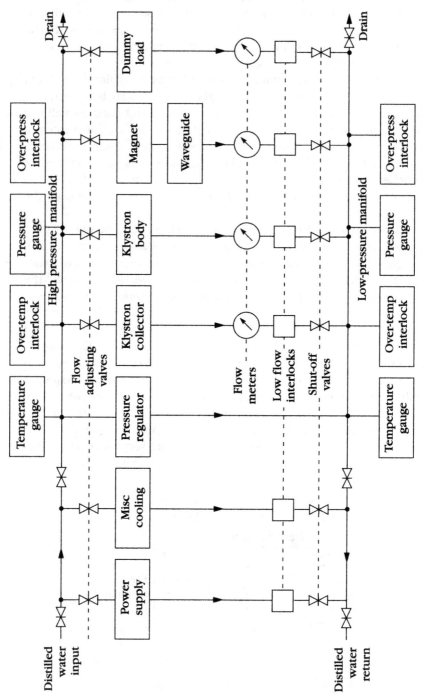

6-2 *Distilled water radar cooling system secondary loop.*

tion. The secondary loop of a cooling system is far more than just water and pipes. The water flow through various parts of the system must be accurately controlled, which requires valves. The addition of valves allows one part of the system to be worked on while the remainder is still operational. Gauges must be provided to accurately track parameters such as water temperature and pressure. Finally, to ensure personnel and equipment safety, various interlocks and sensors are installed.

In this example, six cooling paths constitute the cooling system. Each requires a separate valve to control water flow through it. The more crucial paths will be monitored by individual flow and pressure gauges for maintenance personnel. For equipment protection, temperature and pressure sensors are used. Cooling system parameters such as overpressure, underpressure, overtemperature, and undertemperature must be monitored. If any monitored parameter falls out of specifications, interlocks are provided to shut the radar down. To speed fault isolation, sensors route a signal to the system control panel that illuminates a fault light. As the cooling system often flows around through the high-voltage components of the RF generator, that is the last place that you would want a water leak due to potential shock hazards.

The most common cooling medium is distilled water. To ensure proper operation, certain standards concerning water purity must be maintained. The use of tap water will cause biological growth in the cooling system. Unless it is recommended by the manufacturer, never add any chemicals to control algae and other growth. Although a little chlorine might seem harmless, it could have an adverse effect on RF generator components. If you do experience severe algae growth, the safest remedy is to repeatedly flush the system until it is clear.

Remoting equipment

A radar transmitter and receiver are often installed in one location and the indicators in another. The distances involved can range from a few hundred feet to a few hundred miles. To link the RF section with the indicators calls for specialized equipment. Information that must be sent includes control, status, and radar information. Radar information would be the system triggers, antenna information, and radar video. Status and control information are required to allow for the remote control of the radar system. Radars are equipped with two control panels: the local one installed by the RF equipment for maintenance personnel, and the remote panel mounted by the indicators for the operators. Status information would be things such as transmitter on, high voltage on, antenna scan on, and channel one selected. This type of information would be displayed on both control panels in the form of lights. Control information would be the signals required to operate the radar. They would include items such as high voltage on/off, antenna scan on/off, and STC on/off. The control panel capable of commanding the system is selected at the RF installation.

The shortest distance between the local and remote locations would be on shipboard. Often the two locations are less than 100 feet apart. The usual type of remoting equipment found in shipboard installations is the radar switchboard. If a ship is equipped with multiple radar systems, the switchboard allows each indicator to select a different radar. That allows for maximum efficiency in indicator use. If one is down for repairs, the operator simply changes to a different position. If each display were dedicated to one radar, then the temporary loss of one would impact operations.

Greater distances are involved when the radar is mounted in an aircraft such as the AN/APS-125 long-range air-search radar. The E-2 Hawkeye airborne early warning aircraft will often be located more than 200 miles from the ship, as illustrated in Fig. 6-3. The object is to increase an aircraft carrier's radar horizon. Most shipboard air-search radars have a radar horizon in the range of 200 miles. By mounting a 200-mile-range radar in a high-flying aircraft, the effective radar horizon can be extended out to 400 miles. The data can be shifted to other ships and shore facilities to further increase effectiveness. The data that has to be sent includes radar video, triggers, IFF, and antenna information.

The distances involved in shore-based radars can be up to 10,000 feet. In these instances the remoting system is a subassembly of each individual radar system. Remoting is used so that the radars can be located close to the runways and the indicators located in a control room in the tower. Other remoting installations have been developed that allow the transfer of radar data between Federal Aviation Administration radar installations. Weather radars are also linked together through remoting systems, which allow virtual coast-to-coast radar weather video.

Identification friend or foe

Identification friend or foe (IFF) has been in use since World War II. The idea behind the concept is quite simple: because a radar cannot

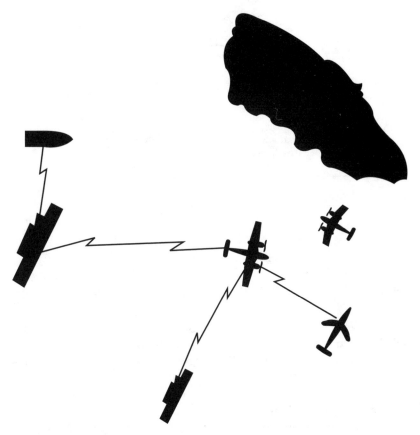

6-3 *Relay of airborne radar data.*

positively identify an aircraft, develop a foolproof method to do so. Through research, a lightweight automatic system was developed that would tag an echo on a radar indicator with coded video. Figure 6-4 depicts the concept. A master set, or interrogator, is triggered by and colocated with the radar system. When the radar transmitter is fired, the interrogator is triggered to transmit. Its output consists of an encoded pulse train radiated along the same bearing as the radar beam. If any ships or aircraft are in the IFF beam pattern, they are illuminated by the interrogator's pulses. If the interrogated target is equipped with an IFF transponder, it will automatically transmit a reply. The reply would be processed and displayed on the radar indicator. As the system is capable of thousands of different codes, air traffic control and military applications are greatly simplified. This widely used and popular equipment is also called *secondary radar.*

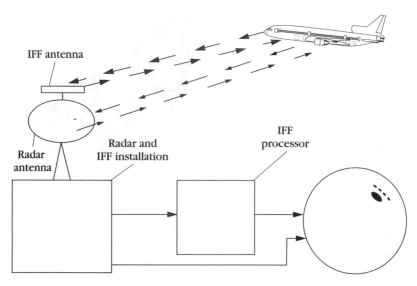

6-4 *Radar and IFF operation.*

IFF has advantages over conventional radar. Because RF energy only has to travel one way, it requires less output power. It is also not susceptible to clutter. Typical IFF ranges extend up to 300 miles. The interrogator is a very small piece of equipment that requires very little space, power, and cooling. Because it is an active system, its responses are not masked by stationary targets and heavy clouds.

It is a simple solution, but it does not completely solve the problem of positive identification. Figure 6-5 is a representation of the early IFF returns. As shown in the drawing, the operator had to interpret the responses. Even with the shortcomings, the system was an improvement over just radar video. Three modes of operation were available, each with its own codes. Additionally, special identification replies were available. Even with separate IFF codes, controllers and operators did become confused as to which aircraft was under surveillance. The special identification feature allows the aircraft to transmit, or squawk, an enhanced code, as depicted in the lower right in Fig. 6-5. Any time an aircraft emergency occurs, ground intervention is required to hopefully avert disaster. With the flip of a switch, an aircraft commander can squawk a coded reply that will alert all air facilities in range that there is a problem. The emergency reply is pictured in the left side of Fig. 6-5. As shown, it is four pulse trains behind the radar echo.

As promising as IFF was at the time, if the system failed for any reason, difficulties arose. Failures could be related to equipment, a

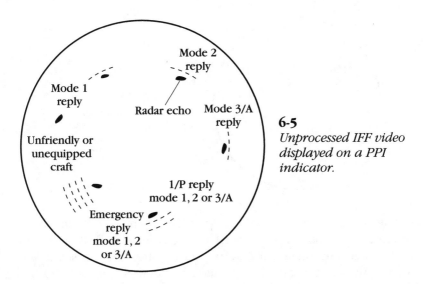

6-5
Unprocessed IFF video displayed on a PPI indicator.

system set to the wrong code, or a system that was simply turned off. This was a problem in both air traffic control and air defense applications. In the 1950s and 1960s, the skies became more crowded with the addition of more commercial and private aircraft complicating air traffic control efforts. The air defense problem was complicated by the fact that aircraft, ships, and ground-based systems could successfully engage targets far beyond visual range. But without a reliable identification method, the unknown target could be hostile or friendly with an inoperative IFF system. The second secondary radar revolution was sparked.

The next step was to develop an IFF system that was more reliable and would improve the tag on the radar display. To complicate the program, any new system had to be compatible with the old one. The obvious solution was to incorporate digital electronics and change the readout on the radar display to an alphanumeric readout. As with the previous IFF systems, the replacement features five modes of operation.

Figure 6-6 is a representation of the resulting IFF codes on a radar display. An IFF-equipped aircraft is identified with an "IX" as a target marker. Notice that the reply has two numeric tags. The upper one is the four-digit IFF code. This is used by all commercial aircraft, suitably equipped private aircraft, and NATO aircraft. The lower numeric code is the altitude readout. In this example, the 155 indicates an altitude of 15,500 feet. If the aircraft has an invalid altitude readout, the second numeric will be ////, the symbol for invalid altitude information. Unknowns are indicated by a circle. An unknown could be an

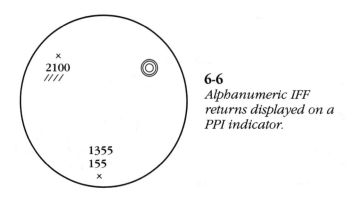

6-6
*Alphanumeric IFF
returns displayed on a
PPI indicator.*

aircraft with an inoperative transponder or an aircraft that is not equipped with one. The system has a special identification feature. To highlight an aircraft, a control will ask the pilot to squawk an "ident." By selecting SPI, a circle marking the reply appears to shrink and expand. If an emergency is in progress, the IFF return will appear to flash, making it stand out on the indicator.

The interrogator operates on a frequency of 1030 MHz with an output power in the range of 1000 watts. The output of the interrogator is a train of three pulses. The distance between the pulses is set by the type of information that is being requested from the transponder. The information the system is capable of providing is identification, altitude, secure military, and special replies. Currently there are 5 modes of IFF operation in use: mode 1, 2, 3/A, 3/C, and 4. Modes 1 and 2 are assigned to the armed forces for tactical identification. Mode 3/A is used by civilian and military air traffic control operations for identification. Altitude information is provided by Mode 3/C. Mode 4 is reserved for military operations as a secure information or coded channel.

Mode operation is determined by the spacing between the first and third interrogation pulses, as illustrated in Fig. 6-7. As shown, Mode 1 spacing is 3 microseconds. The longest is mode 3/C, with a spacing of 21 microseconds. For example, a space of 3 microseconds between the first and third pulses is mode 1, an identification mode. If the time is increased to 21 microseconds, mode C, altitude is the requested information. A and C are the most common IFF modes, for both civilian and military applications. After viewing the following IFF figures, you might want to refer back to Fig. 6-7 to compare the interrogator and transponder codes.

The transponder responses are illustrated in Fig. 6-8, and as can be determined, are more complex than the interrogation pulses. The

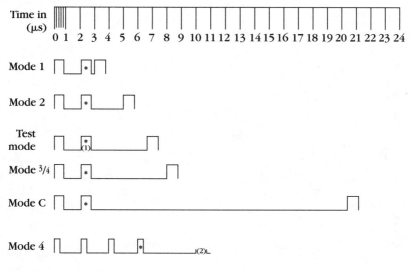

6-7 *Transmitted IFF interrogator pulse trains.*

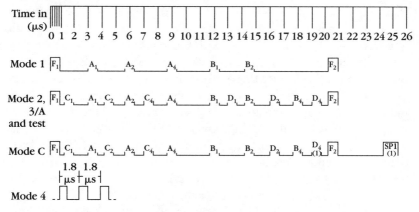

6-8 *IFF transponder replies.*

pulses are all separated by 1 microsecond and are used to modulate a 1090-MHz carrier. Replay information is a little more complex and uses a 12-bit binary code to represent a 4-digit number. Twelve bits are used to give the system a possible 32 codes in mode 1, and 4096 codes are available for modes 2 and 3/A. Mode 3/C, the altitude mode, has a possible 2048 responses. In mode C, or altitude information, height is divided into 100-foot increments.

Special reply functions include the identification of position (I/P), both of which are illustrated in Fig. 6-9. I/P is used when a controller

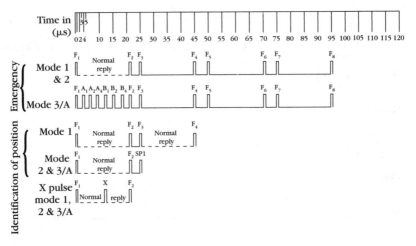

6-9 *Transponder emergency and location replies.*

needs to differentiate between two aircraft. The special function is available in modes 1, 2, and 3/A. In mode 1, the reply pulse train is transmitted twice. Modes 2 and 3/A use only one pulse train, with the addition of an SPI pulse at the end of the signal. The emergency code affects modes 1, 2, and 3/A. As with SPI, there are subtle differences between the modes. In a mode 1 or 2 emergency code, the normal pulse train is followed by three sets of framing pulses. The pulses are 24.65, 44.95, 49.3, 69.6, 73.95, and 94.25 microseconds after the leading pulse. Mode 3/A transmits a 7700 code and the three sets of framing pulses. A conventional IFF transmits at the same rate as the radar it is slaved to. In most instances, interrogations are interlaced. Alternate identification and altitude information are requested from the transponder. The resulting IFF response, regardless of code, is displayed about 40 microseconds after the radar echo.

Figure 6-10 is the IFF installation on board Patrick Air Force Base in Florida. As shown, the unit is very compact. IFF has benefited from the advances in miniaturization, just as other electronic systems have. Due to the small size of the equipment, it can be installed in the same equipment room as the radar cabinets. The IFF antenna, which is also small, is slaved to the radar antenna of the system that provides it with triggers. Figure 6-11 is a typical surface-search radar antenna. In this system, as in the vast majority of installations, the IFF antenna is actually mounted on the radar antenna. In this case the IFF antenna is mounted directly above the radar feed horn. That type of installation saves on space and weight because only one pedestal is required. In the late 1960s there were a few shipboard installations

6-10 *IFF equipment.*

where the IFF antenna was actually mounted separately. They quickly disappeared, so it must have been a failure.

Mapper

As you have already learned, a radar display can be a very confusing picture, consisting of nothing but splotches of video. To ease the demanding jobs of radar operators, a device called a *mapper* was developed. The idea is to superimpose a video map on a radar display that would give the operator a more realistic picture of the area surrounding the radar site.

Air traffic control has become a highly supervised evolution. From the moment an aircraft begins to taxi from the terminal until it unloads at its destination, it is under direct and positive control. Just as the sur-

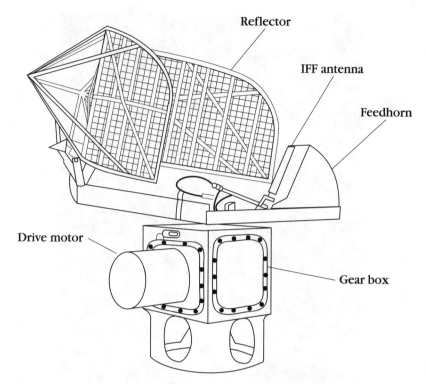

6-11 *Surface-search radar antenna with IFF antenna.*

face of the earth is laced with highways, the sky is an interlocking web of air routes. Every airport is interconnected to others by a mesh of invisible, but nevertheless actual, roadways in the sky. To aid controllers, these routes must be marked on radar indicators that observe them. Topographical features and human-made structures that are hazards to low-flying aircraft must also be marked on radar maps. Other airports and small airfields must also be highlighted, as the aircraft traffic they generate must be taken into account. Restricted zones that preclude air traffic must be readily marked to prevent overflights and incidents. Restricted zones would include military reservations, NASA installations, and important governmental facilities.

The design of a mapper is quite interesting. The basis of a video mapper is a miniature radar display. Figure 6-12 is a block diagram for a representative video mapper. For operation, the unit requires radar trigger and antenna information to synchronize it to the rest of the system. From the radar input, a rotating sweep is developed by the sweep generator and is applied to a small CRT. The CRT will have a diameter of from 2 to 5 inches, depending upon design. A negative

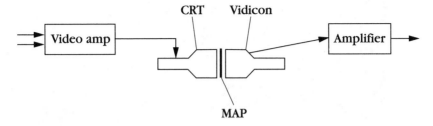

CRT   Vidicon

Video amp

Amplifier

MAP

6-12 *Mapper block diagram.*

of the map is placed over the CRT. As the sweep rotates, it illuminates the map negative. A small video tube, such as a *vidicon,* picks up the video from the CRT. A vidicon is a TV camera tube in which an electron beam scans a charged density pattern that has been formed and stored on the surface of a photoconductor. The photoconductor actually stores the map image, which allows the vidicon to convert it to a video signal suitable for display on a radar indicator. Figure 6-13 is an illustration of a vidicon tube.

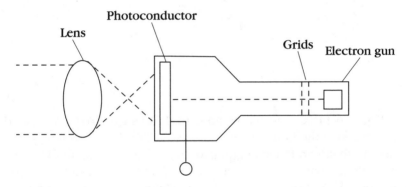

Photoconductor

Lens

Grids   Electron gun

6-13 *Vidicon tube schematic.*

The resulting map video is applied to the radar indicators in the installation. Although early mappers were capable of producing only one map, new equipment can generate up to five simultaneous maps. While this might sound excessive, it does make sense. Often only portions of a map might need to be displayed. An example would be restricted airspace, which is an area that is set aside for specific and controlled flight operations. The White House, military installations, and NASA facilities are examples of restricted airspace. Only authorized and previously cleared aircraft can enter it. If a controller is required to prevent unauthorized traffic, then all the radar indicator needs to display is the outline of the controlled area.

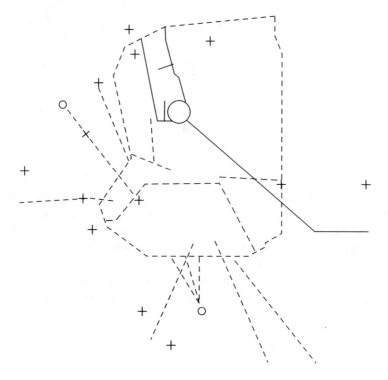

6-14 *Video map.*

Figure 6-14 is an actual video map. In this example, the dashed lines represent the boundaries of the air facility. The solid line, going to the southeast and then bending to the east, is the main traffic route. The outline that resembles a finger is the boundaries of Cape Canaveral Air Station. The four "Ts" are alignment marks for the map, to ensure accuracy. The video map serves the important function of providing the controllers with a permanent reference to ease their demanding job.

Figure 6-15 is a photograph of the FA-7890, the current mapper in wide-scale use by the FAA and the military. This is a mundane-appearing piece of equipment for the major function it handles. This unit is capable of generating five separate video maps. Each radar indicator can individually select which map to display. For the convenience of maintenance personnel, all fuses and controls are mounted on the front panel. A multifunction meter is provided to monitor various equipment functions to aid in failure analysis.

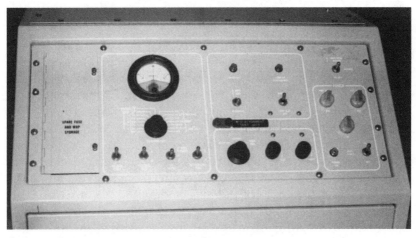

6-15 *Video mapper equipment cabinet.*

BRITE displays

Radar control rooms are considered to be a low-light-level environment. Other than the glow from the CARTS, lighting is limited to small lamps over writing surfaces. The reason is because early radar displays were not bright enough to be used in a brightly lit environment. Most large ships, both civilian and military, have a radar indicator on the bridge. That gives the responsible officer an opportunity to monitor the radar. To use radar in a well-lit location, the radar would be equipped with a hood. The function of the hood is to provide the necessary darkness so that the radar display can be viewed.

The concept of using a radar display in a lighted location was carried over to airport installations. Visual controllers are allowed to control aircraft only if they can be easily seen. In instances of low visibility, it is often difficult to pick out an aircraft. The idea is to give the visual-only controllers access to radar so that aircraft can be identified more quickly. The hood was unsatisfactory because of the constant changes in light. The solution was to develop a television monitor capable of displaying radar video. Although it sounds simple in concept, it was a daunting task.

The outcome was the BRITE family of remote tower displays. The concept is very similar to the mapper. A small CRT receives radar video and antenna information. The CRT displays video, IFF, range marks, and map video. A small TV camera picks up the radar picture on the CRT. Cables route the video signal to the TV monitors. The usual location is in the tower cab, where the visual control of aircraft is accomplished.

Additional equipment includes recorders and emergency power. From the maintenance technician's standpoint, just the knowledge that the emergency power is available is all that is required. The recorders are a different issue. For several reasons, at air traffic control installations and some ship-based systems, audio and video recorders are installed. At airports, all radio communications, interior communications, and radar video are recorded and kept on file for a specified period of time. The archived tapes can aid in accident and incident investigations. If one does occur, the tapes can be played back and viewed on one of the radar indicators. IFF, map, and raw radar video are recorded. In that way, a complete radar picture at the time of the accident can be played back for analysis. Communications are recorded so that there can be no question of the events in the radar control room and the operational condition of the equipment.

Figure 6-16 is a block diagram of an entire radar installation. The system begins with the radar RF section. That would include the transmitter and receiver. External cooling is required to maintain the RF generator temperature within specified limits. The transmitter supplies triggers to the remainder of the system and ancillary equipment. Triggered by the transmitter is the IFF system. Additional triggers are provided to the mapper, BRITE, and indicators. Synchronized by the radar trigger, the mapper develops the video map. IFF, map, and radar video are routed to the remainder of the installation. The BRITE system receives triggers, radar video, and map video. The equipment converts into a format that is suitable for display on a daylight display or television set. The radar indicators receive map, radar, and IFF video for display purposes. The final function is the remoting. All pertinent antenna, video, and trigger information is processed and digitized for long-distance transfer over telephone lines, a microwave link, or RF communications channel.

Some shipboard radar are mounted on stabilized platforms. The idea is to have an automatic mechanical platform compensate for the motion of a ship. Fire control radars and precision approach air traffic control radars are the two types that are most likely to use a platform. These platforms are able to move up to plus or minus 20 degrees in pitch and roll. Pitch is up-and-down motion, while roll is side to side. The main difficulty associated with them is the mechanical alignment.

One final ancillary system is found only with precision approach radars (PAR) on shore bases. A PAR must be capable of providing

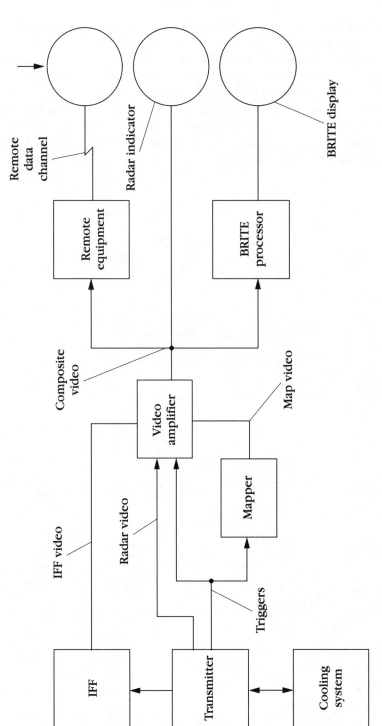

6-16 *Radar installation functional block diagram.*

radar coverage on up to four different runways. The original solution was to use a truck to move a radar system mounted in a van. It has since been superseded by a fully automatic, unmanned mechanical system. Generally very reliable, alignment and lubrication are the main concerns of the maintenance technician.

What makes radar more interesting are the ancillary systems. They allow the radar maintenance technician to do more than just electronics. The cooling systems, mechanical systems, and electronic support systems provide for a change of pace and additional skills.

7

Safety

The knowledge and practice of electrical safety are crucial because radar requires high voltages and current for proper operation. To complicate matters, radar also produces RF radiation, and in some designs, X-rays. To ensure safety and prevent equipment damage, precautions must be known and practiced at all times. Electronic installations require more than just electrical safety. Other hazards encountered by electronics maintenance personnel include: electrical fires, chemical reactions, and working at heights. As you gain experience, you will learn that most, if not all, safety precautions are common-sense actions that can save your life. A good point to remember is that every safety rule was paid for by someone's suffering.

General safety

Before contemplating performing maintenance, proper clothing is a must for any technician. To prevent being caught in moving parts, shirts and pants should not be overly loose or torn. Long-sleeve shirts and pants are preferable to shorts and short-sleeve shirts because long sleeves and pants can provide a degree of protection from burns. There should not be any loose articles in pockets that could fall out and create either electrical or mechanical hazards. It is important that safety shoes, not sandals or athletic shoes, be worn. Radar components and subassemblies can be very large and heavy. Steel-toed shoes prevent dropped objects from causing severe injuries to feet.

Safety glasses or goggles are other, often overlooked, pieces of safety equipment. Electronic maintenance personnel are frequently in areas where eye injury can occur. A good example is replacing defective electronic components. For proper installation, many small components such as resistors, capacitors, and semiconductors must have leads trimmed to fit them into the circuit. When leads are trimmed, small pieces of wire can fly off and strike a person in the eye, causing pain and possible injury. When soldering the component into place, hot solder can spatter, again with the possibility of causing eye damage. When you clean equipment, you use compressed air and solvents, both of which require eye protection.

Electrical safety

A part of electronic maintenance is the proper cleaning and lubrication of equipment. This type of maintenance often has to be performed in a confined space with little or no ventilation. You must exercise extreme caution when using any type of chemical agent. The problem is that chemical fumes displace oxygen, leading to oxygen deprivation, unconsciousness, and ultimately death. Always ensure that you have proper ventilation. If you have any doubts, have trained personnel verify air quality, rather than trusting your nose. Proper safety equipment, such as gloves, goggles, and an apron, might be required to prevent spilling chemicals on yourself. A good idea is to review the material safety data sheet (MSDS) for any new chemical that you encounter. Each chemical used in a work center must have an MSDS on file, which lists all known hazards and safety precautions that are associated with the chemical.

To ensure that equipment cleaning is safe procedure, several steps should be followed. Ensure that the area where the cleaning is being done has sufficient ventilation. If the cleaning agent is flammable, a handy fire extinguisher is a must. To prevent skin and eye contact, wear proper safety clothing, such as rubber gloves, goggles, and a rubber apron. When working with chemical cleaning agents, it's also important that you do not work alone.

The first rule of electrical safety is to never perform maintenance without informing others. The list of possible accidents that could incapacitate you is almost endless. If you are working alone and without others knowing it, it could be hours before you were discovered and rescued. Another good habit is never to work on electronic or mechanical equipment alone. The other person does not have to have any technical knowledge of what you are doing. They just need to be capable of de-energizing the equipment and summoning help. While having two people assigned to one maintenance task might seem extravagant, what is the cost if a person working alone is severely injured or killed?

If you come into contact with an energized electrical circuit, injuries can range from a slight tingling sensation, to burns, unconsciousness, cessation of breathing, ventricular fibrillation, and possibly death. Injuries that result from electrical shock are caused primarily by current flow, not voltage. Even the familiar 115-volt home-lighting circuits can cause injury and death under the right circumstances. If you come into contact with an electrical current, the effects are determined by the intensity of the flow. The smallest current flow that you would feel is about 1 milliampere, and all it would do is produce a slight sensation. An increase to only 10 milliamperes is intense enough to paralyze your muscles. If you had a conductor in your hand, you would be unable to release it. To break contact, outside assistance would be needed. Unless you were removed from the circuit, severe burns and injury would result. An increase in current flow to 100 milliamperes is fatal if the victim is subjected to it for more than one second. Contrary to popular belief, electrocution might not be quick.

The value of current that flows through your body is determined by the applied voltage and your skin resistance. Dry, unbroken skin has a resistance of several hundred thousand ohms and offers protection from shocks. If your skin is damp or cut, resistance drops to less than 500 ohms. The cuts can be as insignificant as paper cuts. If a person with a skin resistance of only 500 ohms comes into contact with an electrical circuit at a potential of 120 volts, a current flow of 240 milliamperes will flow through the body, more than enough to be lethal. Deaths from electrical shock have been recorded when an individual has come into contact with a circuit at a potential of only 30 volts.

To ensure the safety of electrical personnel, first-aid and CPR training should be given to as many people as possible. The first problem in an electrical accident is removing the victim from the circuit. The safest method for both the victim and rescuer is to de-energize the equipment. If that is impossible, use a nonconductor, such as a wooden handle or rope, to remove the victim. Do not under any circumstances allow yourself to become part of the problem by coming into contact with the victim. If that happens, then there will be two casualties instead of one.

To prevent yourself from becoming a statistic, there are several easy-to-follow safety precautions. If you are performing any maintenance that requires opening a cabinet or removing panels to the point where you would be exposed to voltages, tag the equipment out. By placing a tag on an equipment's control panel or breaker, you inform others that a hazardous task is being performed and that they should not energize the equipment or change any settings. A typical equipment installation can have the transmitter in one room, the circuit breakers in a second room, and the operator control panel in a third one. Inadvertent operation is a possibility unless steps are taken.

The presence of a tag is an action that prevents possible unintentional equipment operation while you are working on it. Governmental agencies require adherence to a lockout-tagout program. For civilian organizations, several industrial safety supply houses offer the complete program as a package.

By following the program, chances of injury while performing maintenance is greatly decreased. The steps must be performed completely and in sequence for maximum effectiveness. You begin by ensuring that others are actively informed when maintenance is to be performed on an electronic or mechanical system that they operate. The next step is to place danger tags on all control panels and breakers to warn others not to operate the equipment. At this time, an entry must be made in the lockout-tagout log. To ensure compliance and to warn anyone that might have been inadvertently uninformed, the tags remain in place until the equipment is returned to service. If the system must be de-energized to troubleshoot, replace components, or align, then the control panel, electrical breaker, gas supply, and water supply must be locked out to prevent accidental operation. Again, the placing of a lock on a piece of equipment must be logged to maintain control. You must ensure that all sources of compressed gases, liquids, and power required by the system are secured to prevent operation by using a lock. Only the person who installed the lock is authorized to remove it when repairs are completed. Upon completion of all maintenance, the tags and locks are removed. At that time, suitable log entries are made.

Many people would argue that a lockout-tagout program is more trouble than it is worth. It takes time to clear equipment downtime with all concerned parties and to physically install and remove the tags. From incidents that I have heard about, it is time well spent. One accident occurred in a clothing factory. A corduroy machine needed cleaning, and time to schedule the outage was difficult to obtain. The maintenance person decided on his own to perform the task while the operator was on break. To save even more time, he neglected to tag the machine out, or flip the electrical breaker off, and he crawled inside the machine to clean it. While he was up inside the unit, the operator returned from break early and energized the machine to resume work. His screams were the first sign that the operator had a problem. He was fortunate in that medical assistance was close at hand. His desire to take several shortcuts resulted in several weeks in the hospital due to severe injuries to his legs, and he was let go for violating company policy. The moral of the story is, "Follow safety procedures."

High-powered electronic installations such as radar and communications systems are provided with interlocks as a safety device. The idea is that if access panels are removed, the interlock opens and removes power. That de-energizes the equipment to prevent electrical hazards to personnel. To facilitate maintenance, the interlocks can be bypassed by pulling a shaft out. To accomplish this takes a conscious effort on the part of the maintenance technician. To make maintenance easier, some individuals have been known to bypass or wire around interlocks. The problem with that is that now when the affected panel is removed, power is still present. An unsuspecting person, believing that the interlock would provide protection, is now at risk of electrical shock. Never, ever, for any reason bypass interlocks or other safety devices. Similar to the interlock is the emergency off button. High-powered systems are equipped with the red mushroom buttons in areas where maintenance personnel would be at risk if the system were active. When working on equipment, I use them as a backup to interlocks and breakers. I want to make operation as difficult as possible if I'm the one at risk.

One of the easiest safety precaution to perform, yet one of the most ignored, concerns equipment cabinets. Unless maintenance is in progress, all cabinet doors, access panels, and drawers should be closed and properly fastened. That prevents unauthorized personnel from gaining easy access to the inside of equipment. All nuts, bolts, screws, and zeus fasteners should be in place and tightened. A bad habit exhibited by many technicians is to secure a door or panel with only a few fasteners. Although that makes certain maintenance checks easier, safety is compromised. Normally, it is not a major problem to use one or two fasteners to hold an equipment drawer or cabinet closed. On board a ship, however, the work space is under constant motion due to the action of the ocean. The stresses can cause an energized drawer to spring open, with catastrophic results.

When performing maintenance, it is a good habit not to wear watches or jewelry. They could get caught in moving parts, pulling you into gear trains, shafts, or rotating machinery. Because metal jewelry is a conductor, it is an electrical shock hazard. Necklaces, rings, bracelets, and watches can provide an excellent conductor from an electrical source through you to ground. If you have any items that cannot be removed (such as rings), you should wrap them in electrical tape to provide some protection. To prevent this hazard, I have not worn a watch or any type of jewelry for my entire electronics career.

A good habit to follow when taking any voltage checks is to use only one hand. That way, if you do come into contact with electricity, the path for current flow is through your hand, arm, and then ground. If you place both hands within the equipment, then the path to ground could be through your arm, heart, the other arm, and then ground. By becoming an electrical conductor across your heart, an electrical shock can cause cardiac ventricular fibrillation, or heart failure. Figure 7-1 shows a technician using the proper techniques for working on electronic equipment. Notice that only one hand is inside the equipment where it could come into contact with voltage. As a further safety measure, he is standing away from the equipment so that no other part of his body is touching the cabinet. This ensures that if he does come into contact with voltage, the chances of a fatal shock are greatly lessened.

7-1 *Technician illustrating the proper method of servicing high voltage equipment.* ITT Gilfillan, a unit of ITT Defense and Electronics

Measuring high voltages presents additional hazards. If you place meter leads on energized contacts, you will draw an arc. To prevent injury and equipment damage, there is a safe way to measure high voltages (300 volts or higher). Determine the highest value of voltage that you should encounter in the circuit. Prior to placing your hands inside of equipment, turn the power off and be sure that all components capable of holding a charge have been grounded. Although high-voltage power supplies have safety devices and bleeder resistors to ensure total discharge, never trust them. It is impossible to tell if a capacitor is charged just by looking at it. Therefore, de-energize the equipment and short out the area to discharge any residual electrical charge to ground.

Any shorting must be accomplished only with an approved grounding probe. The probe must be connected to a known-good ground for effectiveness. During the grounding process, touch the end of the probe to any points suspected of having potential present. To make sure you do not become part of the ground path, never come into contact with the probe ground or ground strap. Insert the meter probes, observing polarity. Remove the ground probe, pull your hands from the equipment, re-energize, and take the reading. If the meter scale is too high, DO NOT, repeat DO NOT change scales while the equipment is energized. Meters have been known to break down internally while measuring high voltages if the scale is changed. You must de-energize, change scales, then re-energize. To remove the probes, de-energize the equipment, ground components capable of retaining a charge, then, and only then, remove the meter probes. Any time you are working on de-energized, high-voltage equipment, it is a good practice to always leave a grounding probe in place to ensure safety.

Figure 7-2 is a photograph of a radar transmitter out of an AN/FPN-36 radar. The problem with high-voltage equipment is that energized or

7-2
*AN/FPN-36
transmitter chassis.*
ITT Gilfillan, a unit of ITT Defense and
Electronics

de-energized, it appears the same. The only way that you can be sure that it is de-energized and discharged is through the use of a grounding probe and multimeter. Although the steps seem time-consuming and overly cautious, they must be followed in their entirety. Consider them the minimum level of safety that you will accept while performing maintenance on high-voltage equipment. These measures actually take only a few moments, but they eliminate potential problems.

Electrical fires

Anywhere you have electrical and electronic equipment installed, you have the danger of an electrical fire. Components can short and cause excessive heat, insulating oil can leak, and insulating material can catch fire. All electronics personnel should be familiar with fire safety. The first rule of fire safety is to practice good housekeeping. Properly store all combustible materials, keep the work site clean, keep all pathways clear, and always perform electrical repairs according to specifications. The repair of an equipment with underrated parts can lead to an electrical fire, or at least increased maintenance.

If an electrical fire does happen, the easiest and safest way to fight it is to de-energize the equipment. With the removal of power, most electrical fires quickly go out on their own. If that does not work and the fire is small, use a carbon dioxide extinguisher. Do not try to fight a fire beyond that level. Leave that to the professionals. Fires create a low-oxygen environment that requires breathing apparatus for survival. Burning electronic equipment gives off chemical fumes that can cause injury. So in the event of fire, follow the correct sequence. Turn off the power and secure any ventilation. If the fire persists, try a carbon dioxide fire extinguisher. If the extinguisher fails, or the fire is too large, call emergency personnel. I would only try to put the fire out myself if it were very small. If in trying to fight an electrical fire you were overcome with smoke, who would pass the alarm? In instances such as this, passing the alarm is vital. The paramount consideration is safety. Never take chances of any kind.

There are three classes of fires, based on the type of burning material. Class A fires are combustibles (such as paper, wood, cloth, and trash) that leave an ash. A class B fire is confined to oils, fuels, paints, grease, and materials soaked with any of those substances. An electrical fire is class C and is limited to insulation and combustible materials found in electrical and electronic installations. A fire can start out class C, but, due to poor housekeeping, can rapidly become an out-of-control class A or B.

Due to electrical shock hazards and damage to delicate electronic components and contacts, carbon dioxide is the preferred method of combating an electrical fire. Carbon dioxide does not leave a chemical residue that damages delicate electronics and contacts. Also, it does not conduct electricity, so it is not a safety hazard. Another type of fire extinguisher that can be used on electrical fires is the dry chemical. Potassium bicarbonate is the most common chemical in use as a fire-fighting agent. It is desirable because it is not an electrical conductor, and it quickly smothers flames. However, it does leave a residue that is difficult to remove and can damage equipment. Other common fire-fighting agents are water and foam, both of which are not advisable for electrical fires. The first reason is the shock hazard to emergency personnel. Both foam and water conduct electricity. If either must be used on a fire, the equipment must be de-energized to provide for safety. Unfortunately, water or foam damages equipment, often to the point that it is unrepairable.

Portable equipment electrical safety

A final electrical safety consideration is the proper grounding of all portable electronic equipment and power tools. Normally a three-wire conductor is used to provide power. Two conductors provide power, while the function of the third wire is to ensure that the equipment or tool case is at ground potential. If the equipment is ungrounded for any reason, an electrical potential is present on the case. In effect, the equipment has a floating ground. By coming into contact with the case while grounded, you will provide a convenient path for current flow. As pointed out earlier, a potential of only 30 volts is sufficient to cause injury or death under the right circumstances.

To certify that all portable equipment is electrically safe, an electrical safety program should be instituted. Portable electrical equipment would be portable test equipment, power tools, and appliances. Equipment safety is verified by a few easy-to-perform steps. When checking the ground wire on a standard 120-Vac power plug, make sure that it has a resistance of 1 ohm or less to equipment ground. While performing the resistance test, wiggle the power cord to check for intermittent breaks in the conductor. Visually inspect the power plug and cord, and replace them if they are damaged. Check the equipment case for loose and missing hardware, and tighten and replace as necessary. Whenever replacing a component that comes into contact with ground, ensure that all metallic surfaces are clean and have electrical continuity. Any signs of corrosion must be removed and cleaned. These few steps take very little time to accomplish, but they can prevent injury and equipment damage. The function of any electrical safety program is to verify that the checks are completed on a periodic basis and recorded.

Danger signs

Danger and warning signs are an ever-present part of electronics. Major electronic installations are a hazardous location for the uninformed individual. High-voltage circuits, radiation hazards, ladders, and antenna platforms all have their own threats to safety. To warn of potential dangerous locations, warning signs must be prominently posted and followed.

Any equipment cabinet or equipment that contains high voltage must have a "DANGER HIGH VOLTAGE" sign displayed, as illustrated in Fig. 7-3. Cabinets that have guards or screens to prevent access to high-voltage points should have a danger sign posted inside them. Other signs should warn of radiation hazards from components and antennas. While an antenna is a familiar piece of hardware to those who use it and work on it, it looks harmless to the uninformed.

7-3
High-voltage sign.

There are several different signs, as depicted in Fig. 7-4. The sign in Fig. 7-4A is normally found mounted directly on radar antenna pedestals. Its purpose is to remind maintenance personnel of the dangers associated with RF-emitting antennas. The sign in Fig. 7-4B is a shipboard sign. This type of sign is used to warn nonmaintenance personnel of the fact that RF can be picked up by rigging and metal parts. If the ship has a poor ground, RF burns and shocks are a possibility

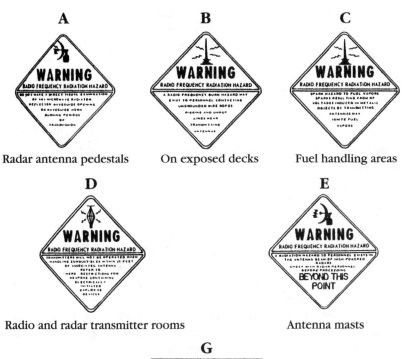

A

WARNING
RADIO FREQUENCY RADIATION HAZARD

Radar antenna pedestals

B

WARNING
RADIO FREQUENCY RADIATION HAZARD

On exposed decks

C

WARNING
RADIO FREQUENCY RADIATION HAZARD

Fuel handling areas

D

WARNING
RADIO FREQUENCY RADIATION HAZARD

Radio and radar transmitter rooms

E

WARNING
RADIO FREQUENCY RADIATION HAZARD

BEYOND THIS
POINT

Antenna masts

G

WARNING
RADIO FREQUENCY RADIATION
HAZARD

On radar control panels

7-4 *Typical radiation warning signs.*

unless safety procedures are followed. RF does present a hazard to flammable materials such as vaporized fuels. To warn of that hazard, the sign illustrated in Fig. 7-4C is used. Even though maintenance and operations personnel are aware of RF radiation hazards, warning signs still must be displayed to serve as a reminder. The sign depicted in Fig. 7-4D is one such sign. This type is typically displayed in radar and communications transmitting spaces. For the safety of all individuals, transmitting antennas are often mounted on masts accessible only by a ladder.

Signs such as the one pictured in Fig. 7-4E are posted at the base of antenna masts and towers. The final RF warning sign is the one drawn in Fig. 7-4G. This type is posted on transmitting control panels, primarily for operations and maintenance personnel. It is usually prominently displayed close to the transmit controls as a reminder of RF hazards. As a competent maintenance technician, some of your responsibilities are to ensure that all signs are mounted, legible, and observed.

Heights

Many electronics technicians must work on equipment installed in high places. As radar and communications antennas must have a clear radiation pattern, they are mounted on towers and masts, often several hundred feet high. Antennas need periodic maintenance and visual examination to ensure problem-free performance. When done correctly, working at heights is a perfectly safe evolution. The problem with working on radar antennas is illustrated in Fig. 7-5. This is an AN/SPN-43 air-search radar antenna installed on board a helicopter carrier. Notice that the work space is very small. Although there is a guard rail, it is not very high. Unless the individuals performing maintenance are tethered with safety harnesses, a fall could easily happen. You can appreciate the height involved by looking at the sailor walking across the deck. Unless safety precautions are followed, this is a very dangerous place.

7-5 *AN/SPN-43 antenna installation.* ITT Gilfillan, a unit of ITT Defense and Electronics

The job begins before climbing a mast or antenna tower. Any maintenance to be performed at heights must be carefully planned, and you must follow ALL safety precautions. First, all equipment capable of transmitting energy must be secured to prevent personal injury. Surprisingly, because of the size and shape of the human body, it can act as a very efficient broadband antenna. Securing all equipment can be time-consuming in the case of shipboard electronic systems. Ships will tie up alongside one another to maximize pier space. That means if you are working on one antenna, another similar one might only be a few feet away. When working in a group of ships, all radiating equipment must be secured to ensure safety.

After obtaining permission, and ensuring that all systems are secured, you can begin to go aloft. Never climb a mast or antenna without a safety observer. In addition to watching you, the observer can warn others to prevent injury if any items are dropped. Once you begin to climb, keep both hands free. If the ladder is equipped with a safety rail, use the appropriate safety harness. Due to moisture, ladder rungs can be very slippery. Any tools that you require should be pulled up with a line. Even the unrestrained motion from a small tool bag can cause you to lose balance while climbing. Another point is that you must tether all tools, parts, and items that could fall. An article that is often overlooked is glasses. For safety, they must be secured with a small piece of string. All tools must be tied or tethered. If you drop a wrench, that means you must climb down, retrieve the tool, and then climb back up to complete the maintenance. Even a small wrench dropped a distance of 100 feet is lethal to any bystanders. While working aloft, do not wear a hat, either soft or hard. Even a small breeze will blow it off of your head. If you try to grab it, you could lose your balance or grip and fall.

Most importantly, you must be tethered. Never work aloft unless you are in a secured area with handrails or are properly wearing an approved safety harness. If you must climb to a location where a harness cannot be used correctly, find another way to safely complete the task. It might require the installation of scaffolding or the use of a crane to ensure personal safety.

Risking your life to finish a job is not worth saving a few minutes. An additional safety problem is present in shipboard antenna installations. Even when tied up to a pier, a ship's power plant might be operational. Any time it is, the stacks emit large amounts of fumes that can be poisonous. The sign illustrated in Fig. 7-6 is used to remind personnel climbing masts of that hazard. By following all applicable safety precautions, working at heights is a safe work environment.

Personnel are cautioned to guard against poisonous effects of smoke pipe gases while servicing equipment aloft.

When servicing equipment in the way of smoke pipe gases use oxygen breathing apparatus and a telephone chest or throat microphone set for communication with others in working party.

Obtain necessary equipment before going aloft.

7-6
Warning sign posted on shipboard antenna masts.

Radiation

High-powered radar and communications systems are hazardous because of the presence of high voltages, currents, and RF energy. The RF energy radiated by an antenna can induce voltages into ungrounded metal objects such as antenna towers, ladders, handrails, and wire guys. By coming into contact with an ungrounded metal structure, individuals can receive a shock or RF burn. The painful burns, if severe enough, can lead to death.

RF is such a hazard because the human body acts as a broadband antenna due to the length of trunk, arms, and legs. If an individual is exposed to a high-energy RF field, the effects might not be noticeable for hours, or even days. Because microwave energy is the cause, it is the same as if microwaving food. When a human being is exposed to a high-powered RF source, the result is called *dielectric heating*. Dielectric heating occurs when an insulating material is placed in the presence of an RF field. The heating is caused by internal losses that are from the rapid reversal of polarization of the molecules composing the dielectric material in the human body. Any resulting damage is caused by heating from inside out. As the interior of the human body has relatively few nerve endings, the effects are not initially noticeable. The product is often very deep and penetrating third-degree burns that can frequently be as small as a pinhole.

I know the effects of RF burns from firsthand experience. Early in my training, I was practicing maintenance on an IFF transmitter. To complete the task, the antenna had to be disconnected. Due to haste, I neglected to reconnect the antenna, I energized the system, and I transmitted 1000 watts at a frequency of 1030 MHz into my hand. The very small burn looked as if it had been caused by a pin or tack. Although the burn was very painful and took several weeks to heal, it was a very cheap lesson. This accident, like most, was caused by several events: inattention to detail, careless placement of my hand, not following procedures, and working alone.

When exposed to RF radiation, the eyes are the organs frequently affected first. Exposure can lead to cataracts and possible blindness. For safety's sake, never look into RF-producing devices or stand in the path of RF-radiating devices. This safety rule dates from the early days of radar research. Several researchers viewed the interior of waveguides and RF-producing devices to see if any visible phenomena were produced. Tragically, the result was that some of them went blind. The reason why we know of the risks is because of what happened to them. Never forget that radiation damage is cumulative. It might take years for you to absorb enough RF radiation to suffer noticeable damage. To ensure that you do not have future health problems, never expose yourself.

RF energy is lethal if absorbed in large enough quantities. The U.S. government has computed the lethal distances for virtually all high-powered, RF-producing equipment. To minimize risks, access to antennae and transmitter rooms is limited to prevent accidental exposure. On shipboard antenna installations, red lines mark the closest point that one can approach an antenna without harm. Other, more dangerous systems have locked ladders and access points to prevent unauthorized entry. This is important because radar can often cause physical injury and death in distances measured in the hundreds of feet. The hazard associated with a particular radar system is determined by the frequency, power out, and beam pattern. The amount of RF energy that you are exposed to is determined by your distance from the antenna and the length of time spent in the beam pattern. The longer your are in a beam, the greater the level of your exposure. The farther away from an RF-producing source you are, the less RF energy is absorbed by your body. The radiation density decreases by a factor of the square of the distance. Remember, distance is your only protection from a radiating antenna. There is no safe way for you to be close to a radiating antenna. More importantly, there is no reason for you to be close to one.

The amount of RF energy produced in a given location by a system is determined by the antenna motion. A stationary beam is far more dangerous than a rotating one. When the beam is stationary, the transmitted RF energy is concentrated in a very small area. As the beam is stationary, the amount of time you receive radiation is greatly increased. A moving beam, such as a rotating radar, is much safer. The amount of time exposed is greatly decreased because even a slow rotating antenna will have four or more rotations per minute. A rotating beam means that you will be exposed for only a few seconds. Even a moving beam emits RF energy, so be aware of the hazard and act accordingly.

RF energy is also a hazard to inanimate objects. It has been known to set off explosive devices such as explosive bolts and proximity fuses. Large fields have also been known to ignite combustibles such as gasoline. Ignition happens because of the heating and arcing that can result. Due to the hazard, there are areas on an aircraft carrier's flight deck where fuel tankers and gasoline-powered vehicles cannot be parked due to RF hazards.

Electron tubes

Electronic tubes present several hazards to the uninitiated. However, with information and a few precautions, there is nothing to worry about. First, if you are replacing an electron tube, never use your bare hands. If the equipment has been operated recently, a tube could still be very hot. Just like metal, it is impossible to tell if a tube is hot or cold. To safely remove it, always use either a tube puller or appropriate heat-resistant gloves. I recommend welder's gloves or asbestos gloves because both have superior heat resistance.

Another problem associated with tubes is that some of them contain radioactive material. Typical radioactive tubes include TR, ATR, spark gap, gas-switching, and cold cathode tubes. While not all of them contain radioactive material, some do in significant quantities. Tubes containing radioactive material are required to be appropriately marked.

Defective tubes should be disposed of in accordance with EPA and OSHA regulations. As long as the tube envelope remains intact, no hazards exist. If for any reason the envelope is ruptured, radiation and radioactive material can escape. To ensure personnel and environmental safety, several precautions must be followed. Radioactive tubes should remain in approved cartons until installed in equipment. Defective tubes should be placed in cartons to prevent damage. Never place a radioactive tube in your pocket for safekeeping. It could be broken there, complicating cleanup. If a tube is broken, then notify supervision and follow all procedures for cleanup and disposal.

Cathode-ray tubes (CRTs) present their own set of hazards due to their sheer size and internal vacuum. The internal vacuum places almost two tons of force on the surface of a 10-inch CRT. If the glass envelope is ruptured in any manner, implosion results. The resultant force turns the internal CRT components into missile hazards. Any time a CRT is removed, proper handling and disposal instructions must be followed.

Due to the internal pressures, never scratch or strike the surface of a CRT. When removing or installing one, do not force it. Any excessive force could snap the neck, resulting in implosion. When working around a CRT, ensure that the equipment is de-energized and that the high voltage is discharged. Do not ever hold a CRT by its neck. The structure is not that strong and can snap. Handle a CRT gently, as any rough handling can cause internal damage. When setting a CRT on a surface, place it face down on a thick piece of soft material to prevent screen damage. Whenever handling a CRT, wear safety glasses, gloves, and an apron. CRTs must be disposed of in an appropriate manner due to the implosion risk and the face coating. The internal face of a CRT is coated with a chemical phosphor that is toxic. Prior to disposal, the tube must be rendered harmless by releasing the internal vacuum. The safest way is to place the defective tube in a carton face down. Find the locating pin mounted in the base. Figure 7-7 is the pin diagram of a typical CRT. The locating pin is the largest and the center pin. Using a pair of pliers, very carefully break the pin. The removal of the pin allows air to enter the tube.

In this chapter, radar and other RF-producing equipment sound dangerous. If approached correctly, however, you have nothing to fear. First, know your equipment. Does it have a lethal zone you cannot enter? Know the values of voltage used in every circuit. Be aware

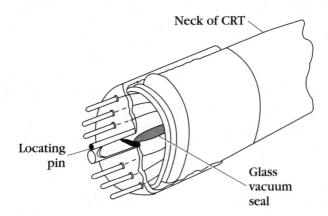

7-7 *CRT base construction.*

of hazards such as large-power-supply capacitors and pulse-forming networks. Use that knowledge in conjunction with all safety practices. Know the rules and follow them. Never take chances, and never take shortcuts. The few moments saved are not worth the cost if you have an accident. Good safety practice begins and ends with a pilot's phrase, "situational awareness." Always be aware of where you are and what you are doing. As safety is ultimately your responsibility, electronics operation and maintenance are as safe as you make them.

8

General maintenance considerations

All electronic equipment maintenance can be broken down into two broad categories: preventive maintenance and corrective maintenance.

Preventive maintenance

Preventive maintenance is any maintenance task that is performed on operational equipment with the purpose of verifying operation and maintaining peak performance. Preventive maintenance can range from simple observation to alignments and adjustments. Corrective maintenance would be any maintenance task performed to correct an equipment deficiency and return it to peak operational performance. As an electronics maintenance technician, you can expect to perform more preventive maintenance than corrective maintenance on a major electronics installation. The effectiveness of any radar system hinges on the type of maintenance and attention that it is given. A misaligned receiver reduces the amount of information available for the rest of the system, even if the RF generator and indicators are perfectly adjusted. To ensure peak performance, the entire system must be properly maintained.

General-purpose test equipment

Before performing any maintenance, an electronics technician must be familiar with general-purpose test equipment. Both corrective and preventive maintenance require the use of several different test instruments. The better a technician understands test equipment, the easier the job will be.

One of the most common, versatile, and important pieces of test equipment is the oscilloscope. A simple, representative oscilloscope is pictured in Fig. 8-1. The function of an oscilloscope is to give a visual presentation of circuit action. Through the use of this test instrument, a technician can observe and measure frequency, waveform duration, phase relationships between signals, waveform shape, and amplitude.

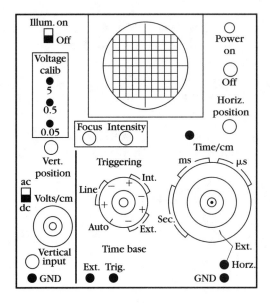

8-1
Simple representative oscilloscope.

There are literally scores of different types of o'scopes in use today. Basically, they all function the same. The horizontal axis of the CRT is time (such as seconds, milliseconds, and microseconds). The vertical axis is amplitude in volts. To provide for more stable viewing, the instrument is usually triggered by the system under test. Radar pretrigger is a good timing waveform to use for the synchronization of external test equipment. A time control is used to set the duration of the horizontal sweep. Most oscilloscopes are dual channel. That means that two separate signals can be observed at the same time and compared. At one radar site, I was fortunate enough to have a five-channel test instrument, which made troubleshooting and alignments easier and quicker. Today, many o'scopes are microprocessor controlled, with a multitude of special features in a small package. As a radar technician, you will use oscilloscopes for the alignment and failure analysis of radar systems.

Multimeters are also important test instruments. Currently, all meters fall into two broad categories: digital multimeter and analog meter. At a minimum, a good multimeter should be capable of measuring ac/dc voltage, dc current, and resistance. Additional and desirable features are continuity tests and ac current. *Continuity* is a position that has the meter beep if a conductor under test is a short circuit. An innovation of the 1980s was the digital multimeter (DMM), which has virtually taken over the multimeter field. DMMs are extremely versatile test instruments capable of performing a wide range of tests. Some of the high-end DMMs have features such as temperature, frequency, capacitance, and inductance measurements.

As flexible as these talented instruments are, there are still some tests performed better by the old-style analog meters. I have found that an analog meter is superior when testing the changing resistance of a potentiometer. Also, when performing antenna alignments on a precision-approach radar, the analog meter has a better indication of circuit action. However, with the rapid pace of change in electronic test equipment, the days of the analog meter are numbered.

The signal generator has many applications in radar maintenance. Due to the different characteristics associated with frequencies ranging from audio to RF, signal generators are classified by frequency range. A simplified signal generator is pictured in Fig. 8-2. To properly maintain a radar, you probably use two separate instruments: one

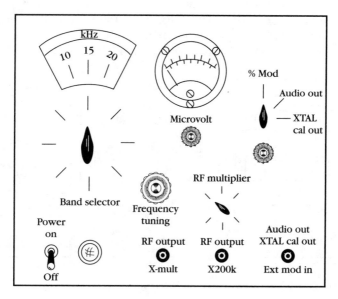

8-2 *Simple representative signal generator.*

for IF frequencies and one for RF frequencies. The units will be used to inject known frequencies for test and alignment purposes.

A variation of a signal generator is the range mark generator. This is a specialized piece of test equipment developed for the proper alignment of radar indicators. The instrument produces a series of video pulses that appear at fixed distances. The video pulses are then used to accurately align the indicator's range strobe and range marks.

A *spectrum analyzer* is a highly specialized piece of test equipment that is capable of observing the frequency spectrum produced by an RF generator, such as a magnetron or klystron. The unit features a CRT display to give the technician a visual indication of the transmitter's output. By periodically testing the frequency spectrum produced by a radar, you will have an indication of transmitter performance. With this test instrument, component problems and misadjustments that are normally very difficult to analyze are quickly isolated. Due to the high voltages and currents present in the transmitter of a radar, locating problems is challenging. However, with a spectrum analyzer, failures related to the RF generator and modulator are unmistakable.

A *time delay reflectometer*, as pictured in Fig. 8-3, is a highly specialized piece of test equipment that is vital when needed. The test instrument is used to troubleshoot failed cables, wiring harnesses, and conductors that have shorts, opens, crimps, or defective connectors. If an antenna has a coaxial feed, then it can be used to check for defective cables. As a technician, you have to check many cables and wires with continuity tests. While that type of investigation is satisfactory at low frequencies and short distances, there are times when it is inadequate. Rather than just testing for continuity, which is only a resistance check, a TDR uses radar basics.

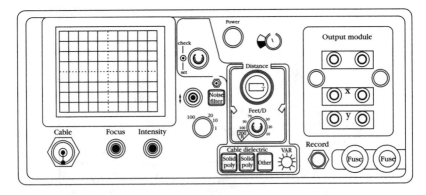

8-3 *Time delay reflectometer front panel.*

A TDR functions by transmitting a short burst of energy into a cable under test. The instrument is then used to observe for reflections at the point of signal insertion. If no energy is reflected back to the point of insertion, then the cable has a uniform impedance and it is functional. Any defect in the cable is reflected back as a discontinuity, which can be positive, negative, or fast changing. The position of the discontinuity on the CRT display indicates where in the cable run the fault is located. When testing 10,000-foot coaxial cables used to route radar information under runways, it is vital to know exactly where the fault is located. By using a TDR, you can determine that there is a short/open in the cable 4000 feet from the radar site. By using system drawings, the problem is isolated to a specific access point, facilitating rapid repairs. The unit can be just as useful on shipboard installation where cable runs can be several hundred feet through several different decks and bulkheads.

Although preventive maintenance is vital to the successful operation of any system, it is often misunderstood and ignored. As with safety precautions, maintenance procedures were developed in response to real problems and issues. It has been discovered over the years that timely preventive maintenance saves many times its costs in reduced corrective maintenance work hours, parts, and equipment downtime.

One very important preventive maintenance step that is often overlooked or ignored is equipment cleanliness. Many potential equipment problems are eliminated by simple cleaning and good housekeeping standards. If grease or oils are allowed to accumulate on electronic equipment, large amounts of dust are attracted and allowed to build up. Far more than unsightly, this can lead to equipment failures. The dust forms a high resistance path to ground for current flow. The existence of these paths leads to component failure and equipment downtime. Accumulated dust causes components and subassemblies to run hotter. Heat is an enemy of sustained equipment operation. Heat can directly lead to component failure. The hotter a part operates, the shorter its operational life. Even if component failure does not result, overheated components can shift in value, changing circuit parameters. Locating a failed component in a circuit full of components changed in value is a challenge. Dust buildup in high-voltage circuits can lead to arc overs. Once a "trail" has been burned into an insulated surface, it must be replaced.

Air filters are an important preventive maintenance check. To cool electronic equipment, many systems require an unimpeded air flow. To minimize dust problems, filters are fitted to clean the moving air. The filters are important because if they become blocked with dirt, air flow is decreased, increasing equipment internal temperature and leading to equipment downtime. Equipment air filters come in two broad classes: disposable and nondisposable. If the disposable filter is dirty, just replace it. If a nondisposable filter is dirty, it must be cleaned. The best method is to clean it with soapy water. After all the dirt and dust are removed, rinse it in clear water. Then, time permitting, let it air dry. If time is a problem, use low-pressure air to blow the moisture out.

Equipment cooling systems are another area that requires periodic attention. Most cooling systems are equipped with meters to track important system parameters such as pressure and temperature. By periodically recording the values, system operation is tracked. If a parameter changes, that indicates the need for more in-depth maintenance before damage results.

Cooling systems have filters to remove large impurities. If a filter is blocked or only has partial flow, this increases equipment temperature and leads to component failure. To minimize cooling system failures, tap water should never be used in them. Water suitable for human consumption contains minerals that can cause deposits in small channels and in the filter system. RF cooling systems often have small channels for water that pass through the heat-producing components. Also, tap water can lead to biological growth, such as bacteria and algae. If biological growth is a problem, flush the system until the water is clean. You should never add chemical agents unless it is recommended by the manufacturer.

As a minimum, on performing preventive maintenance you follow what the manufacturer recommends. With experience, you might find that local conditions change the frequency of some checks. A good example is air filters. Typically, an air filter is a monthly check. One radar site that I was assigned to was so dusty that radar indicator air filters had to be changed on a weekly basis, rather than monthly.

System checks

Alignments and adjustments must also be verified and performed on a repetitive basis. Many times when a piece of equipment is to be used for a demanding operation, that type of maintenance must be performed more frequently. Verification of equipment accuracy and operational checks are performed on a daily, weekly, monthly, and quarterly basis. Although this seems to be time-consuming, by adhering to the schedule, actual effort is reduced and equipment availability is increased.

The most recognizable repetitive radar check is ring time. Ring time is considered to be the benchmark test of a radar system's operation. To perform the test, an echo box is required. Figure 8-4 is the block diagram of a radar system with an echo box installed. A directional coupler is used to connect the echo box to the radar system. The idea is to operate a radar normally while it is under test. The echo box is actually a tunable resonant cavity used as a test instrument. Its function is to capture a portion of the transmitted RF energy and, during receive time, release it back into the radar system. The captured energy causes the echo box to oscillate, or ring. Due to its resonant characteristics, the echo box rings for a short time after the transmitted pulse. The energy the echo box returns to the system is processed by the receiver as a normal return. Ring time is a rough check of system power output, receiver sensitivity, and frequency. The result is either viewed on an oscilloscope monitoring the output of the receiver or on the radar display. Figure 8-5 is an illustration of how ring time appears on an oscilloscope. Notice that it appears as a large pulse. By viewing it with an oscilloscope, pulse duration and fall time can be verified. The pulse duration is directly determined by the output power of the transmitter, receiver sensitivity, and system frequency. Pulse fall time is directly attributable to duplexer recovery time.

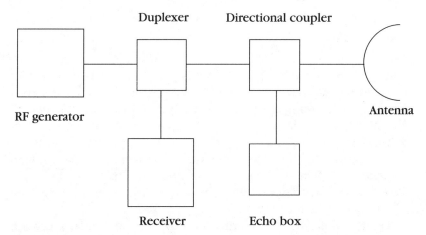

8-4 *Radar system block diagram showing echo box.*

Figure 8-6 is a sketch of the internal construction of an echo box. The tuning mechanism actually changes the size of the resonant cavity, and cavity size determines the resonant frequency of the test in-

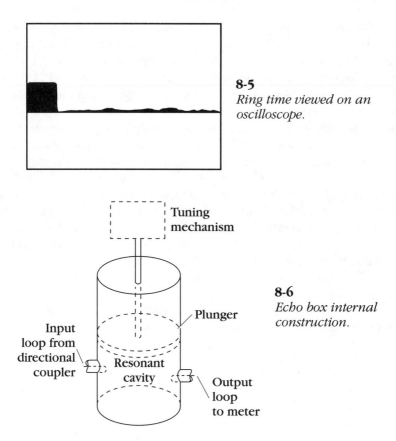

8-5
Ring time viewed on an oscilloscope.

Tuning mechanism

8-6
Echo box internal construction.

Plunger

Input loop from directional coupler

Resonant cavity

Output loop to meter

strument. The output from the echo box is rectified and applied to an analog meter. A peak on the meter indicates the resonant frequency of the box. The tuning mechanism is equipped with a direct read dial. When the meter peaks, you have an accurate frequency measurement of the RF generator. To determine the radar's frequency, the tunable cavity is provided with a frequency dial, as shown in Fig. 8-7. System frequency is obtained by viewing the ring time on an oscilloscope. The echo box cavity is varied to obtain the maximum ring time. The radar frequency is read from the echo box frequency dial.

Figure 8-8 illustrates how ring time appears on a radar indicator. Viewing it on a radar indicator is preferable because it verifies the operation of the remoting, cables, video amplifiers, and radar indicators as well. Because the ring time on a radar system can extend out 4000 yards or more, it can be taken only with the approval of the operators. To maintain peak system operation, ring time is normally taken several times a day.

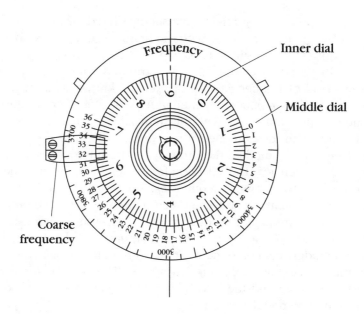

8-7 *Echo box frequency dial.*

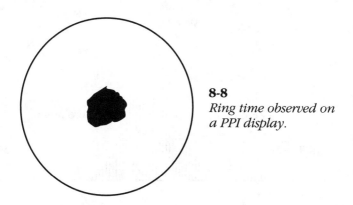

8-8
Ring time observed on a PPI display.

RF generator performance tests

A very important transmitter check is RF generator output frequency. For optimum system performance, an RF generator is designed to operate within a very limited band of frequencies. If the transmitter is operating outside of that frequency band for any reason, system efficiency is degraded, and interference with other radar is possible. RF generators can be either fixed frequency or tunable. If a fixed-frequency radar is off frequency, the most likely cause is a defective magnetron. A tunable radar that is off frequency could be caused by a defective RF-generating component, or it might have simply drifted off frequency. Before replacing any parts, try tuning the unit first. If it still exhibits frequency drift, then investigate replacing the RF generator.

The RF output of a transmitter is distributed symmetrically over a narrow band of frequencies. The pattern is known as a *frequency spectrum.* An RF generator operating at peak efficiency produces a frequency spectrum of known and predictable configuration. Figure 8-9 illustrates an ideal frequency spectrum. When one frequency is used to modulate another, the product is a base frequency with two sideband frequencies. The sideband frequencies are the sum and difference of the two original frequencies. The result is that the RF output of a pulse radar consists of numerous frequencies. The output is actually carrier frequency modulated by short pulses occurring at the PRF of the radar. By analyzing the RF output of a radar, it is evident that two modulating components are present in the signal. One is the PRF and its naturally occurring harmonics. The other is the fundamental frequency and the odd harmonic frequencies that make up the rectangular modulation pulse.

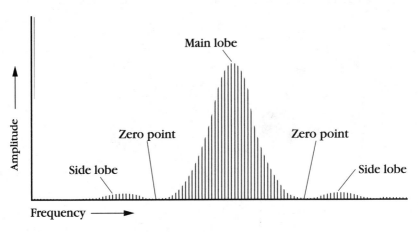

8-9 *Ideal magnetron frequency spectrum.*

By referring to Fig. 8-9, the concept of the carrier and modulation frequency interrelationship can be viewed. The vertical lines in the drawing represent the modulation frequencies produced by the PRF and its harmonics. Notice the two lobes on either side of the main lobe. The main lobe represents the modulation frequencies produced by the carrier frequency and its associated harmonics. The side lobes are produced by the odd harmonics of the carrier frequency. The zero points between the lobes are the result of the even harmonics of the carrier frequency. In an ideal radar spectrum, the frequencies above the center frequency are a mirror image of those below.

The following four figures compare several different radar frequency spectrums. The spectrum in Fig. 8-10 represents an RF generator operating with design specifications. Notice that the side lobes are a mirror image of each other, well-defined and perfectly shaped. The side lobes are separated from the main lobe by perfect zero points. While the transmitter output illustrated in Fig. 8-11 lacks the well-defined side lobes, it is still functional. This could be considered to be the beginning of degraded RF generator performance. The spectrum in Fig. 8-12 indicates that the transmitter is in need of maintenance. The lack of zero points above and below the center frequency of the main lobe is evidence that the transmitter output is being frequency modulated during pulse time. A possible cause for this defect could be in the modulator function. A modulation pulse with sloping rise and fall times or lacking a flat peak is one strong possibility. The RF generator itself could also be the cause if it were unstable or the permanent magnet were weak. If the magnet were the problem, the most common symptom would be high magnetron current and drifting AFC. Other potential causes could be a high-voltage power supply with a low output or insufficient tube current to the generator.

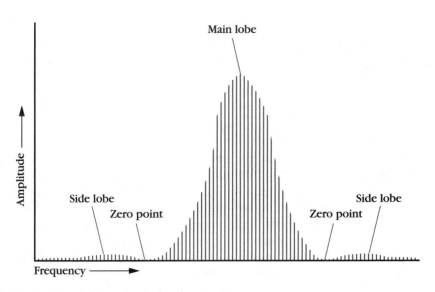

8-10 *Normal transmitted pulse spectrum.*

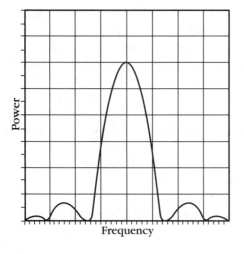

8-11
Fair transmitted pulse spectrum.

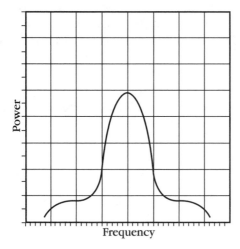

8-12
Poor transmitted pulse spectrum.

The frequency spectrum in Fig. 8-13 is of an obviously defective radar transmitter. Notice the lack of zero points and side lobes. An irregular spectrum is indicative of severe frequency modulation. Due to the rapid frequency shifts, the receiver AFC is unable to accurately track, which leads to greatly reduced system effectiveness. The first place to check would be in the adjustment of any tuning stubs. If they are properly adjusted, the magnetron is probably defective. Figure 8-14 illustrates an RF generator that is double moding. *Double moding* occurs when an RF tube rapidly shifts between two frequencies. This type of problem is the result of either a faulty tube or standing waves. Standing waves can be caused by a defective antenna rotating joint, an obstruction in the waveguide, or improper connections.

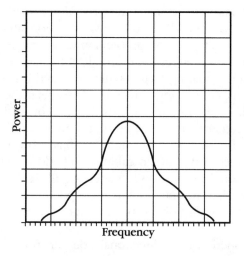

8-13
Defective transmitted pulse spectrum.

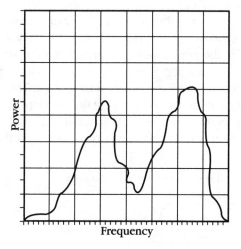

8-14
Double moding.

The use of a spectrum analyzer can isolate failures to the RF generator. Additional checks are used to verify that the RF generator is defective. The most common symptom of most magnetron problems is low magnetron current. When you observe the radar display, background noise or grass will appear to be normal, but any displayed echoes will be weak, possibly indistinct, and fuzzy. AFC will be erratic, drifting, or inoperative. Testing the system with an echo box will indicate a very short ring time.

RF generator output power is crucial to system operation. The high output power and RF frequencies require specific techniques to measure transmitted power. Power measurements are taken by sampling the output power. Most radar is equipped with a directional coupling in the waveguide system to simplify the procedure. Readings from the power meter are taken in dBm and converted into watts using a conversion chart. The most common problem in obtaining an accurate power measurement is forgetting to add in all of the losses. A directional coupler has an attenuating factor, which is noted on a permanently attached tag. Any connecting cables used between the coupler and power meter also have inherent losses. Ideally you will use calibrated cables. If not, charts are available that convert cable type and length into losses. To calculate transmitter average power, just add up all losses (the directional coupler, cables), and add them to the reading on the front panel of the power meter. As all figures are in dBm, use a conversion chart to obtain an average power reading. To obtain a peak power reading, divide the average power by the duty cycle of the radar.

If the radar is not equipped with a directional coupler, power measurements are somewhat more difficult. You must use a pickup horn in front of the antenna to obtain a sample of the transmitted power. In this case, in addition to the cable and horn losses, you must factor in losses caused by the atmosphere, distance from the antenna, and weather conditions. The reading is once again taken with a power meter, and the value of dBm must be converted into watts. This test is not as convenient or as accurate as using a directional coupler.

There is a very important reason for an accurate transmitter frequency. Figure 8-15 is a representative comparison between the fre-

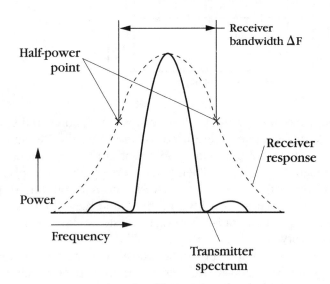

8-15 *Receiver bandwidth compared to an RF generator frequency spectrum.*

quency response of a typical radar receiver and the output spectrum of a representative radar transmitter. Notice that the receiver has a wider bandwidth than the transmitter's output pulse. That ensures complete reception of any returned energy. However, the best receiver response is in the center of the curve. From the center frequency, receiver response tapers off rapidly. Adequate receiver response is available between the two half-power points. For maximum system efficiency, the center frequency of the receiver and transmitter should be aligned.

Although not a preventive maintenance check, a new magnetron requires special handling to ensure a long service life. When replacing a magnetron, it must be seasoned, or baked in, for proper operation and a long operational life. Internal arcs are very common in new tubes and are caused by the liberation of gases from the tube elements. Internal arcs can also be caused by sharp surfaces inside the tube, mode shifting, and operating at excessively high currents. The cathode structure can withstand arcs for a short time. However, damage can result, destroying the tube. The seasoning process is to slowly increase the tube voltage until an arc occurs, burning off liberated gases. If arcing is persistent, lower the voltage. Do not resume increasing voltage until the tube settles down. Seasoning is complete when operating levels of voltage and current have been reached and the tube functions without arcing. Proper seasoning might take several hours.

Receiver performance tests

Efficient system performance requires a receiver that is aligned and operating to design specifications. The most important receiver maintenance checks are receiver sensitivity, receiver bandwidth, and TR recovery time. Many radar are equipped with special circuits to improve system operation, such as IAGC, STC, FTC, and AFC. Unless the manufacturer requires it, all alignments are accomplished with these special functions not selected.

Receiver sensitivity has a direct impact on maximum detection range of targets. A decrease in receiver sensitivity has the same influence on detection range as decreased transmitter output power. If a receiver has a 3-dB decrease in sensitivity, it will have the same effect on range as a 3-dB decrease in output power. While that change in output power is easy to detect, it is not in the receiver. A simple misadjustment is more than enough to induce a 3-dB loss. Figure 8-16 compares overall system performance to maximum achievable range. For example, if both the receiver and transmitter have a decrease of

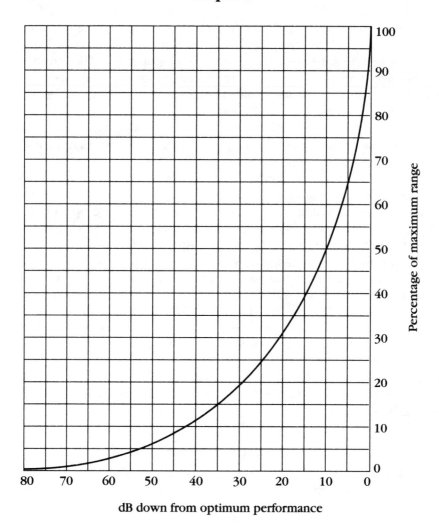

dB down from optimum performance

8-16 *Performance versus range.*

10 dBm from maximum, that translates into a 20-dBm drop. A change in 20 dBm results in a maximum range of only 30% of optimum. If the radar has a normal maximum detection range of 200 miles, then a 20-dBm reduction results in a range of only 60 miles.

Receiver sensitivity is a gauge of its capability to detect weak echoes. The test of a receiver's sensitivity is called minimum discernible signal (MDS). MDS measures the power level of the weakest signal that a receiver can detect and is illustrated in Fig. 8-17. An MDS is the smallest signal that is capable of producing a visible receiver output as observed by an oscilloscope or on a radar indicator. To

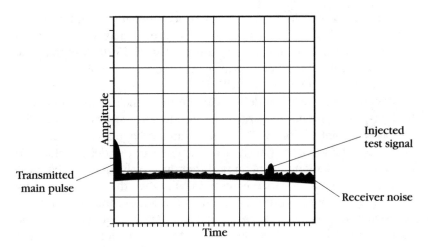

8-17 *MDS viewed on an oscilloscope.*

maintain peak system performance, MDS checks should be taken on a periodic basis to track system performance. For accuracy, each separate test must be performed identically.

The first step is to ensure that the receiver is accurately tuned and properly tracks the transmitter frequency. If there is a difference between the receiver and transmitter frequencies, then that indicates either a tuning problem or an equipment failure.

To perform the MDS test, a signal generator is used to inject a known test pulse signal into the directional coupler. There are two current test methods. The pulse technique will be discussed first. The signal generator inputs a fixed frequency test pulse. The output of the receiver is monitored with an oscilloscope. As with the transmitter power test, take into account all losses, including the cables and coupler. While observing the test pulse on the oscilloscope, slowly increase the signal generator's output attenuator. Stop increasing the attenuator setting as soon as the test pulse disappears into the receiver background noise, or grass. Decrease attenuation until the test pulse just becomes visible. The sum of all attenuations is equal to the receiver MDS. To accurately perform this test takes practice. Each time you do the test, repeat it several times to ensure accuracy.

The second method uses an FM frequency generator to input a frequency that sweeps across the bandpass of the receiver. The test is performed exactly as with the pulse signal generator. This test is more accurate because it tests receiver sensitivity across the entire receiver bandpass.

Receiver sensitivity is always stated as a negative number and is referenced to 1 milliwatt. Therefore, if a receiver has a sensitivity of –75 dBm, that means that it can detect a signal 75 dB less than the 1-milliwatt reference level. Modern radar receivers typically have a sensitivity in the range of –105 dBm.

Receiver bandwidth is defined as the frequencies found between the half-power points of a receiver frequency response curve. The half-power point is also 70.7% of maximum voltage level. Each radar has its own bandwidth, which is determined in the design phase. If during tests and alignments you find that the bandwidth has changed, then corrective maintenance is required. Bandwidth is determined by injecting a test pulse in the directional coupler. Set the generator on the center frequency. Increase signal generator frequency until the pulse decreases by 3 dB. Repeat the test on the lower side of center frequency until the lower half-power point is found. Subtract the lower half-power point frequency from the upper half-power point frequency. You now have the receiver bandwidth.

The final receiver test is TR tube recovery time. TR recovery time is vital because it controls the minimum range of a radar. Until the TR deionizes, reconnecting the receiver to the antenna system, it is incapable of picking up any returned RF energy. Many manufacturers state that recovery time is how long it takes receiver sensitivity to return to within 3 dB of normal levels after the transmitter fires an output pulse. Depending on design, recovery time can range from 1 microsecond to 20 microseconds.

The function of a TR tube is to protect the input receiver circuits from the high-powered transmitter pulse. All tubes allow some transmitted energy to leak into the receiver front end. However, the amount of power is so small that it has a negligible effect. With age, though, TR tube leakage increases until it reaches a level that burns out the receiver crystals. Some radar technicians have developed a practice of replacing TR and ATR tubes after a certain number of operational hours. While this is somewhat effective, it is not a complete answer. A better solution is to periodically check and record TR tube performance. Each radar has the accepted TR tube recovery time tests listed as a part of the technical manual.

Another effective but easy-to-perform test is to monitor tube keep-alive current. The function of keep-alive current is to maintain a tube in a partially ionized state. That allows a tube to ionize faster, better protecting receiver circuits. A good TR tube normally draws 100 microamperes of keep-alive current. As a tube ages and loses effectiveness, keep-alive current decreases.

Waveguide and antenna checks

Antennae and waveguides are considered to be low-maintenance sections of the radar installation. Performance checks are usually limited to periodically inspecting an antenna system. Rotating antennae must have periodic oil changes and have moving parts lubricated. Failure to maintain the proper oil level will result in a defective antenna drive system. During the periodic inspection, the material condition of the antenna reflector, pedestal, and motor must be checked. Exposure to weather can deteriorate metal components and electrical cables.

Antenna alignment is crucial in radar systems used for precision purposes. To maintain a high degree of accuracy, alignments are performed frequently. Antenna position is converted into a voltage level by special circuitry. To perform the test, the antenna must be manually moved through its range, and its outer voltage limits must be set. As a verification, the voltage generated by the midpoints is checked against known standards.

Part of the inspection of other types of radar is to verify antenna alignment. All rotational antennae must be referenced to a fixed point to ensure bearing accuracy. On shore-based systems, the pedestal has a true north benchmark engraved on the pedestal. The antenna is stopped and manually aligned to the benchmark. To perform this adjustment, the system must be in standby for personnel safety. With the antenna aligned to true north, the stationary sweep on all radar indicators should be pointing north. If any are not, then the indicators need to be aligned.

Shipboard radar have to be aligned to both north and the ship's bow. The antenna has a microswitch aligned with the bow. When the antenna passes or stops on that point, a *ship's head* is generated. A ship's head is a video blip appearing on the radar indicators that is aligned with the ship's bow. For true north, the antenna is aligned to the ship's gyrocompass. A switching arrangement is provided so that the operator can select either true north or relative as the antenna's reference point.

Waveguide periodic maintenance is primarily limited to ensuring that it is painted to protect from corrosion. All joints should be inspected to verify sealing. An improperly sealed joint can allow moisture to enter the system.

One test performed on antennae and waveguides is the standing-wave measurement test. A standing-wave measurement test can tell a technician a great deal about the RF transmission function of a radar system. The presence of excessive standing waves in a waveguide indicates an impedance mismatch between the receiver, transmitter, and antenna system. If a mismatch is in the system, RF energy is reflected from the load back to the source, or RF generator. Standing waves induce an inefficiency into the RF section that degrades system performance.

The location of a standing-wave measurement is very important. The reflected energy from a mismatch varies in phase the same way that the propagated energy does. At certain points, the propagated energy and the reflected energy are in phase. The points are located one-half wavelength apart. At that point, measured voltage is maximum. One-quarter wavelength away, the two energies are out of phase, resulting in a minimum voltage measurement. The ratio of the maximum to minimum voltages is called the voltage standing-wave ratio (VSWR). The ideal VSWR is expressed as 1:1. That indicates that the RF system has a good impedance match, and maximum energy is propagated. A high VSWR indicates the presence of an impedance mismatch, which results in power losses. The maintenance manual for each radar will describe how the VSWR check is to be performed.

A low VSWR is desirable because it indicates that the maximum amount of energy is being propagated through the waveguide system and the antenna. Mismatches are undesirable because the reflections cause improper transmitter operation, such as double moding. A high VSWR can also lead to arc overs in the waveguide system. If that happens, then the affected sections must be replaced.

Shore-based precision approach radar are units that are usually located alongside the runway. Because airfields have multiple runways, a method had to be provided to allow the system to be moved. In the early days of ground-controlled-approach radar, the systems were installed in wheeled trailers and moved with a truck. This less-than-satisfactory method was replaced by a fully automatic turntable. Designed to be operated unmanned, the radar changes position with the push of a button. The mechanics of the turntable system requires both preventive and corrective maintenance. The primary preventive maintenance checks are lubrication, inspection for corrosion, and alignment. The most common corrective problems are the drive motor and chain. One aspect that makes radar maintenance more interesting is the addition of ancillary equipment such as the cooling systems and turntable.

Cooling system checks

For the most part, radar cooling systems require very little maintenance. Maintenance is limited to periodically checking the water and filters in the secondary system for cleanliness. Cooling water in the replenishment tank should be clear, without any scum or debris floating on the surface. If the water becomes cloudy or murky, the only solution is to flush the system until it is clear. Additives such as bleach or chlorine should not be used unless authorized by the manufacturer. The water filter should be checked on at least a quarterly basis for buildup. Local conditions might require a more frequent observation. One check that never seems to be mentioned is the condition of all valves in the system. They should be checked to ensure that they are not frozen and that they turn freely. Any emergency overflow drains should also be kept clear and free-running.

If you ever have to perform maintenance on the RF generator, it might be necessary to drain the secondary cooling system. It is difficult to remove all of the water from the system. If the radar is to be shut down for an extended period of time, it is imperative that all moisture be removed to prevent biological growth. The easiest way to accomplish this task is to use low pressure air and a wet/dry vacuum cleaner. Begin by de-energizing the radar and draining as much water as possible. After that, remove the source and return lines. You then use the wet/dry vacuum cleaner to draw the water out of the system by placing the hose over the drain port. At the same time, apply a very low-pressure air on the source port. After water stops flowing, reverse the air and vacuum connections. This will almost completely dry the secondary cooling system.

Radar indicator tests

Radar indicator tests can be summed up as alignments, alignments, and alignments. The radar display is the one point where everyone judges the efficiency of a radar system. For optimum system operation, the displays must be accurately adjusted. The most common adjustments are range ring accuracy, range strobe accuracy, sweep linearity, cursor linearity, and display shape. Display shape is for appearance; the others are for accuracy.

For proper viewing, a radar display must be circular. While an egg-shaped display does not affect accuracy, it can be distracting. For best results, follow the manufacturer's procedures. Circularity was not much of a problem until manufacturers began building indicators with flat-faced CRTs. Properly aligning flat-face correction circuits can be very time-consuming.

To accurately perform radar indicator alignments, a calibrated range-mark generator is required. The test equipment generates accurate range marks at various ranges. To perform the test, simply inject the output of the range-mark generator into the indicator's video input. For convenience, the range-mark generator also produces its own radar triggers. Therefore, the performance of an indicator can be tested without using any signals from the remainder of the radar system. The position of the internally generated range marks is checked against that of the ones produced by the test instrument. The indicator's range-mark generator has a position adjustment that allows you to position the marks on top of one another.

Range strobe accuracy is also verified using the range-mark generator. Using the external range marks, position the strobe over one. The range strobe digital readout should agree with the positional value of the selected range mark. If it does not, set the range strobe to the position where the digital readout agrees with the external range mark. The range strobe generator board will have a position adjustment. Using the adjustment, move the range strobe until it lines up with the selected range mark.

Curser and sweep linearity adjustments are performed using an oscilloscope. Because each radar indicator is vastly different, follow manufacturer recommendations. Other radar adjustments include internal video levels. They are more or less set where the operators need them for local conditions.

The most important component in a radar indicator is the cathode-ray tube. Whether the system is operational or turned off, it attracts attention. Most radar CRTs have burned spots in the center. This happens either during installation or from a failed sweep. An intensified spot strikes the face of the tube rather than the sweep. To eliminate this problem, always turn intensities down when aligning a CRT. Most modern indicators are equipped with self-protection circuits that turn off the high voltage if the sweep fails for any reason. While that saves the CRT, it does complicate failure analysis.

Corrective maintenance

The first step to correcting a failure in any electronic system is knowing how it appears when it is operating normally. The time to learn how a system operates is before any failures occur. To fully understand it, it is advisable to use an oscilloscope to observe all system waveforms. By observing and recording correct input and output waveforms for each chassis, you have a starting point when isolating failures. But, before investigating system operation through waveform analysis, you should know your equipment.

Figure 8-18 shows a modern radar installation, the AN/SPS-48E three-dimensional air-search radar. As can be seen from the photo-

8-18 *AN/SPS-48E equipment installation.* ITT Gilfillan, a unit of ITT Defense and Electronics

graph, equipment space is limited, and the cabinets could be confusing to an untrained person. You must accurately know the exact location of each function. Valuable troubleshooting time is lost if you are frantically searching for the low-voltage power supply or the MTI receiver. The knowledge of actual system layout is very important.

More problems than you realize are caused by lack of familiarity with the radar system's control panels. That problem is called "knobology." Both maintenance and operational personnel need to be familiar with the proper setting of both local and remote-control panels. Many times a control in the wrong position can give the appearance of a failed system.

As shown in Fig. 8-19, a control panel has numerous buttons, lights, and switches. This example is the local control panel for an AN/GPN-27 airport surveillance radar system. You must be aware of

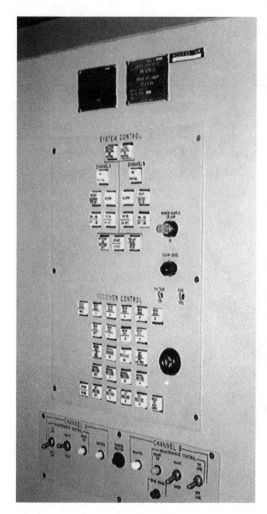

8-19
*Airport surveillance
radar local control panel.*

what each control does and its proper setting in each mode of operation. The resulting status lights must be just as familiar to you.

Knobology directly relates to "KISS," the acronym for "keep it simple, stupid." Always check the simple things first. If the complaint is that the antenna is not rotating, make sure that the antenna control is in the scan or on position. No MTI video or a lack of range marks on an indicator could be an intensity turned down.

Failure analysis always begins with observation. Before doing anything, just observe. If you are in the radar control room, what is the visual condition of the radar indicators? How is the control panel set up? Which panel lights are on, and which ones are off? Through careful observation, most failures can be isolated to one or two functions, just from front-panel indications and controls.

Troubleshooting an inoperative piece of equipment should be accomplished logically. Approached correctly, the time required for corrective maintenance can be greatly reduced. The easiest way to troubleshoot electronic equipment is called the six-step procedure. The logical six steps are:

1 Symptom recognition.
2 Symptom elaboration.
3 Identifying possible faulty functions.
4 Localizing the faulty function.
5 Localizing the faulty circuit.
6 Identifying the failed component.

When I was in Memphis as a Navy radar instructor, we had several problems that drove home the need for a logical approach to troubleshooting. One of the more interesting ones involved the radar antenna. To simulate a fault, we disabled the antenna on command in the remoting equipment. The result was that when the system was placed in remote control, there was no antenna motion. When the equipment was placed in local control, the antenna would scan normally. Many students failed to pick up on the effect on system symptoms when placing the unit in local control. Very few of them noticed that in remote control, the antenna was turned off, but it functioned normally in local control. When students followed the six-step procedure, this was a 10-minute problem. If they failed to, then they never found the defective component. As a result, students referred to this problem as "the L.A. freeway with no off ramps."

Symptom recognition is vital because in order to repair a piece of equipment you must first know if it is functioning normally. A piece of equipment is designed to perform a specific function at all times. If it fails in any way, then it is in need of maintenance. A trouble symptom is a sign of a malfunction in an electronic system. Symptom recognition is the action of recognizing a malfunction. Symptoms can be obvious or very subtle. That is why it is imperative that you are familiar with the normal operational characteristics of equipment you are maintaining.

The easiest symptom to recognize is equipment failure. When failure occurs, a major function or the entire system is not performing. A good example would be the loss of main power. When that happens, the entire system or equipment is not functioning. Panel lights are out, cooling fans are inoperative, CRTs are dark, or the antenna is not rotating. Degraded performance is a little more difficult to ascertain. Low transmitter power or a decreased receiver sensitivity are examples of that type of problem. The equipment is still nominally functional, but something does not seem right because distant targets are difficult to pick up. In order to perform this step, you must know the normal operational capabilities of your equipment.

Symptom elaboration is the second step. At this point, you are still observing the operation of the equipment. Is one indicator blank, or are all five of them blank? In the case of the distance problem, are all transmitter and receiver front-panel meters normal? This is the step where you use front-panel meters and built-in test equipment to verify system operation. This is the point where you check control panel switches. Is the antenna defective, or is the antenna scan switch just turned to the off position? By carefully following this step, you can narrow the possible faulty functions to only one or two chassis within the radar system.

The third step is the action of identifying a list of possible faulty functions. This is a direct result of how well the first two steps were performed. If all symptoms were correctly identified, then the situation is progressing well. This step is applicable to equipment that contains more than one function or operation. Examples of functions within a radar would be power supplies, triggers, modulator, receiver, remoting, and indicators. If a radar indicator were the equipment under repair, then the functions would be power supplies, trigger generator, sweep generator, range-mark generator, video amplifier, range strobe, and cursor generator.

The first three steps did not rely on the use of test equipment. Rather, you used your senses and knowledge of the equipment to narrow the likely area of the failure to as few circuits as possible. In the fourth step, you determine which function is at fault. This is the first step where you actually use test equipment to check equipment parameters. There are three factors to consider in this step. Choose measurements that will eliminate as many functional units as possible. Another point is the accessibility of test points. If the unit must be disassembled to check a test point, then you might want to skip it at this time. The final consideration is past experience with the equipment. If you have seen the exact symptoms before, then that might lead you directly to the problem circuit.

The fifth step is localizing the failed circuit. How you isolate the failed circuit depends on the arrangement of circuits within the function. Figure 8-20 illustrates several possibilities. If the failed function has 20 different circuits connected in series, that it is known as a *linear path*. Because of the number of circuits, you would want to min-

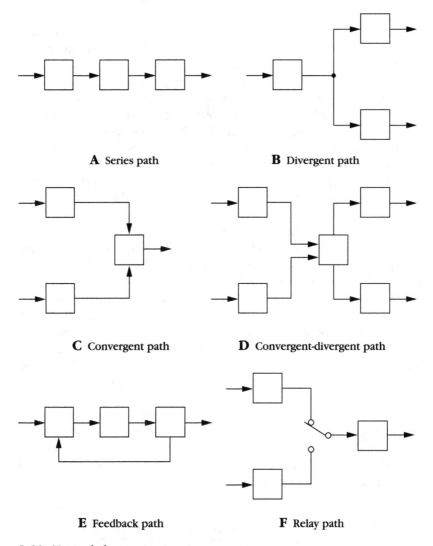

A Series path

B Divergent path

C Convergent path

D Convergent-divergent path

E Feedback path

F Relay path

8-20 *Typical electronic circuit arrangements.*

imize the number of checks that you have to take. The most logical approach is to split the number of circuits under test in half. If the output of number 10 is good, then you know that the failure is between 11 and 20. The next check would be to split the circuits in half again by checking number 15. In this manner, only a few checks should isolate the failed circuit, rather than 20, saving a great deal of time. This approach is called *bracketing.*

If the function has a divergent path, then there is no most-logical first check. In cases such as this, personal preference would determine where you began your observations. In a convergent path arrangement, begin with the output circuit. In the convergent-divergent arrangement, begin with the common circuit. The most challenging is the feedback path. In this type of layout, check the input on the first stage. This type of problem is the most time-consuming. To find the exact circuit, you must break the feedback loop where it is routed to the input stage. The final path is the switching path. This example is nothing more than a relay circuit. A point to remember is that the output of one circuit might go to a circuit located in a different chassis or circuit board. Cables, wires, circuit board pins, and connectors can fail. Just because the output of one circuit is normal, that does not mean that the input of the following circuit is good.

After you have definitely identified the failed circuit, you are ready for step six. In this step, the goal is to find the faulty component and review the steps you took to get there. You will use detailed schematics and test equipment. At this point you should be checking the operation of a single circuit consisting of an active device and support components. An active component would be a transistor, vacuum tube, or integrated circuit. You start with voltage checks. That tells you if the device is saturated or cut off. The voltage checks will lead you to the component that has caused the problem.

Review is an important part of the troubleshooting process. Through review you learn how to improve your failure analysis skills so that you can become a better electronics technician. Table 8-1 is a review of the six-step troubleshooting method. Each step lists its goal and which tools you are to use in its completion. Troubleshooting is as much a science as it is an art.

General maintenance

In radar maintenance there are general points that pertain to all systems. To ease maintenance tasks, you should have a set of nonmagnetic tools for working around RF generators equipped with permanent magnets. It is a challenge to properly tighten bolts when the magnet keeps pulling your hand away.

I have found it helpful to maintain my own set of schematic diagrams. That allows you to personalize them with your own notes. Another aid that I use is equipment history. If you compile a history of every failure that an electronic system experiences, you develop a powerful troubleshooting aid. With a written history, you will find

Table 8-1
The logical six-step troubleshooting method

Step 1	Symptom recognition	Eyes
		Ears
		Knowledge
Step 2	Symptom elaboration	Eyes
		Ears
		Knowledge
		Controls
		Notes
Step 3	Possible faulty functions	Eyes
		Ears
		Knowledge
		Notes
		Equipment diagrams
		Think
Step 4	Localize faulty function	Eyes
		Ears
		Knowledge
		Notes
		Think
		Equipment diagrams
		Test equipment
Step 5	Localize failed circuit	Eyes
		Ears
		Knowledge
		Notes
		Think
		Equipment diagrams
		Test equipment
Step 6	Localize failed component	Eyes
		Ears
		Knowledge
		Notes
		Think
		Equipment diagrams
		Test equipment
		Review and learn

that many problems are repetitive. It also helps with the rare but difficult problems by giving others a starting point.

Spare parts are an area that you should be very interested in. A good rule of thumb is to maintain enough parts to last for one year. All electronic components have a finite shelf life. Components such as capacitors might last only five years. A defective spare part is worse than having no part. A bad habit to get into is board and component swapping. If you suspect a circuit board, but the spare does not cure the problem, reinstall the original board. If you are fortunate enough to have a spare set of boards, verify that they are operational by installing them in the equipment. After verification, store them in the original packing material in an approved cabinet.

Modern components require proper installation techniques. Attending a soldering course will save you time, effort, and mistakes. Also, many components today are static-sensitive. As such, they must have special handling, storage, and installation steps to ensure normal operation.

Figure 8-21 is a subassembly from an AN/SPS-48E. You will find that most circuit boards today are tightly packed. Anytime you are taking waveform or voltage checks, exercise extreme caution. The slip of a test probe can short out two leads, destroying an integrated circuit or transistor. Another example is the subassembly from an AN/FPN-36,

8-21 *Radar equipment chassis open, illustrating circuit components.* ITT Gilfillan, a unit of ITT Defense and Electronics

8-22
Open equipment drawers call for caution. ITT Gilfillan, a unit of ITT Defense and Electronics

illustrated in Fig. 8-22. In this example, the circuit board is mounted on a card extender to facilitate circuit analysis with test equipment. If you look in the lower left, you will see a rail. Unless the assembly is locked in place, it can easily slide into the equipment, causing damage. This is an all-too-often occurrence. A technician will forget and will close the drawer with the circuit board still in the extended position. Now, instead of a failed component, the board is destroyed.

Summary

A valid question is, "Does radar have a future?" The answer is yes. To some individuals, advances in stealth technology have erroneously promised to render many military applications obsolete. However, recent articles in military journals indicate that large-scale stealth technology might have already been rendered obsolete by the simple expediency of resurrecting older, lower-frequency radar. The lower-frequency units can detect stealth aircraft because the baffles that deflect high-frequency RF are invisible to them. If anything, radar will continue to evolve and play an important role in our lives.

NASA is currently flying missions using the space shuttle as a platform for the latest imaging radar, the SIR-C. The function of this system is to provide multifrequency, multipolarization imagery of the planet earth. The 16,000-pound system consists of two separate radar systems that provide researchers with images and measurement of vegetation, water storage, ocean dynamics, wave fields, wind fields, volcanoes, tectonic activity, soil erosion, and desertification. Each 10- or 11-day mission provides volumes of crucial data.

Military applications are even more numerous. Figure 8-23 shows a U.S. Navy cruiser. In the center of the ship are two circular antennae, one above another. These are missile-control radar. Looking higher is the black square antenna that is the external sign of the AN/SPS-48E three-dimensional radar. If you look carefully on top of the 48 antenna, you can see the horizontal bar that is the IFF antenna.

8-23 *Modern warships are dependent upon numerous radar systems.* ITT Gilfillan, a unit of ITT Defense and Electronics

Directly above it is the surface-search radar antenna. In the lower-left side of the picture is the Close-In-Weapons-System. The white dome on top of it is its tracking radar. If anything, military vessels will become more dependent on radar technology.

Civilian applications are also increasing. More and more pleasure craft and small boats are equipped with radar. Air traffic control, medical electronics, and industrial flaw detection will also increase the number of radar systems in use. Just a few years ago, a well-equipped airport had only an airport surveillance radar. Today, IFF, the surveillance radar, wind shear detector, weather radar, and ground-traffic radar are required to ensure safe operation. Modern radar systems are becoming so compact and light that they can now be used to locate underground utilities. Radar control is spreading from aircraft to ships. Governmental agencies in many countries use shore-based radar systems to monitor water-borne traffic to reduce accidents and groundings. In medical electronics, radar technology is used to combat cancer and detect various diseases. Radar will be a useful tool for humans for many years to come.

The first radars were temperamental electronic mazes that required almost daily maintenance and attention. Reliability increased with improved circuitry and designs. True breakthrough in mean time between failures (MTBF) began with transistorization. Systems ran cooler, decreasing downtime for several reasons. Tubes are a high-failure item, with a relatively short operational life span. To provide for compact radar cabinets, tubes were mounted on their sides. The problem with horizontal mounting is that the delicate internal metal structures of the vacuum tubes sagged, increasing failures. Transistors ended that. The advent of integrated circuits brought a quantum increase in reliability. It is now common for radar to operate hundreds of hours without any electronic or mechanical failures. Now computerization and electronic scans promise to further improve the reliability. In addition to the reliability, radars now have special features that make each individual system more valuable. For example, the AN/GPN-27, which the FAA calls the ASR-9, combines the features of an airport surveillance radar and weather radar. Any failures must be corrected quickly and accurately. There is a downside to this for technicians. You gain experience only by working on equipment. To improve your skills and maximize troubleshooting time, always review system theory, and constantly train. Through familiarity and review, you will hone your technical skills. (See Fig. 8-24.)

As you have learned in this book, radar is both fascinating and complex. It takes effort and dedication to learn how a system oper-

8-24 *The reward for extensive training and hard work is a reliable, trusted radar system.* ITT Gilfillan, a unit of ITT Defense and Electronics

ates and how to maintain it. The end result of the effort is a radar system that an operator uses with confidence under all conditions. To be an effective radar technician, you must be well-versed in all aspects of this fast-paced and expanding field. By virtue of the wide range of equipment that comprises a total radar system, every day is different and challenging. Radar system maintenance encompasses virtually every facet of modern electronics technology and can be one of the most rewarding technical career fields today.

Appendix

Abbreviations

AGC	automatic gain control
ASR	airport surveillance radar
CCA	carrier-controlled approach
CRT	cathode-ray tube
CW	continuous wave
dB	decibel
EPI	exponential planned indicator
FM	frequency modulation
GCA	ground-controlled approach
FTC	fast time constant
IC	integrated circuit
IF	intermediate frequency
LO	local oscillator
MHz	megahertz
MTBF	mean time between failures
MTI	moving target indicator
MW	megawatt
PFN	pulse-forming network
PPI	planned position indicator
PRF	pulse repetition frequency
PRR	pulse repetition rate
PRT	pulse repetition time
STC	sensitivity time constant
TWT	traveling wave tube

Glossary

A scope A radar indicator capable of displaying range information only. Bearing data is read off of a digital readout. Similar in appearance to an oscilloscope display.

airborne radar Radar systems installed in aircraft for navigation, weather, altimeters, weapons control, and collision avoidance.

airfield surface movement indicator Low-powered radar installed at major airports designed to track aircraft and vehicles on the ground.

air search radar Radar designed to track airborne targets from near ground level to high altitudes. Provides a 360 degree coverage.

airport surveillance radar An air search radar optimized to track aircraft within sixty nautical miles of an airport. Typical coverage from ground level to 30,000 feet.

altitude Vertical distance of an object above a known reference, such as the earth's surface.

ambiguous An incorrect or illogical indication. An example would be an ambiguous target range—an incorrect target range.

antenna gain The ratio of the signal strength of a particular antenna as compared to a reference antenna.

antitransmitted receiver (ATR) tube A gas-filled tube that electrically disconnects the transmitter from the antenna system during receive time.

approach control radar A relatively short-ranged radar that is designed to track aircraft approaching an airfield or aircraft carrier. Must be capable of providing accurate target information in regions congested with clutter.

asymmetrical multivibrator A multivibrator that produces rectangular output pulses.

automatic gain control Circuit designed to compensate for targets of widely varying amounts of returned energy.

average power The average value of RF power radiated by a radar transmitter during one complete cycle of operation.

azimuth An angular positional measurement in the horizontal plane as compared to a reference point.

balanced mixer Located in the input section of a receiver. It is a waveguide arrangement that has the shape of a T. It uses two crystals connected to a balanced transformer to cancel out local oscillator and input stage noise.

band A continuous range of frequencies.

beacon An RF transmission, stable in frequency and intelligence, that is used for navigation and direction finding.

bearing resolution The ability of a radar system to differentiate between two targets at the same distance, but different bearings from the antenna.

bistatic A radar system that has two separate antennas; one for transmission and one for reception.

blind speed The velocity at which an aircraft directly approaching the antenna system of an MTI-equipped radar will be invisible.

blip A term used to describe the reflected energy from a target displayed on a radar display.

broadside array An antenna in which the direction of the maximum radiation is perpendicular to the vertical plane of the array. Also called a bedspring.

bulk effect An electrical effect such as current, resistance or resistivity as observed in the entire body of a material rather than just the surface or a junction. As an example, a diode junction exhibits resistance and bulk resistance. Resistance would be just that of the PN junction and bulk resistance would be the junction plus all the possible paths around it.

carrier-controlled approach radar A shipboard radar system that is designed to guide aircraft to landings on aircraft carriers during periods of reduced visibility. The information can be used for advice, or as a fully automatic landing.

cavity A metallic chamber that allows RF energy to reflect. Resonance can occur if the cavity physical dimensions are correct.

characteristic impedance The value of equivalent resistance, that when used to terminate a transmission line results in no reflections of RF energy.

circulator A multiport microwave coupler that allows the controlled transfer of RF energy from one port to another.

clutter Unwanted echoes that interfere with the ability of a radar system to observe targets of interest. Clutter includes vegetation, topographical features, structures, and the surface of bodies of water.

coherent A phase relationship between two electrical signals or waveforms, such as the RF energy transmitted and received by a radar system.

contact An object that reflects RF energy in sufficient levels to be displayed on a radar indicator.

continuous wave modulation (CW) A periodic waveform resembling a sine wave that is not interrupted in any way.

copper loss Power loss that is associated with metal conductors.

cosecant-squared antenna A radar antenna that produces a cosecant-squared beam pattern.

cosecant-squared beam A radar beam pattern that has an intensity that varies directly with the square of the cosecant of the angle of beam elevation.

cross section, radar (RCS) The effective area of an object that is capable of reflecting RF energy back to a radar system. RCS is determined by the target's shape, construction materials, and angle to the radar.

dielectric A nonconducting material that is used as an insulator.

dielectric loss The rate of the transformation of electrical energy from a changing field into heat loss.

diode detector A demodulator that used semiconductor diodes to rectify the IF signal into a video signal suitable for use on a radar indicator.

directivity The ability of a radar antenna to transmit and receive energy more efficiently in certain directions as compared to others. It is also the degree of beam sharpness produced by an antenna system.

discrimination The ability of a radar to accurately track a target in a region of clutter.

discriminator A circuit in which phase or frequency variations are converted into amplitude variations.

Doppler effect The change in frequency of received RF caused by the relative motion between the radar system and a moving object.

duplexer The electronic switch that is used to switch the antenna between the receiver and transmitter. In the process, the duplexer is to protect the input circuits of the receiver from the high-powered transmitter pulse.

duty cycle The ratio of active to inactive time of a radar transmitter.

echo box A resonant cavity that is used to measure the performance of the radar transmitter and receiver. A small portion of the transmitted pulse is retained by the echo box and returned to the receiver.

echo, radar RF energy reflected back to the antenna system by an object.

electric field The free space surrounding a charged body or conductor in which the electric energy can have an effect.

electronic scanning A method of providing antenna motion through the use of electronics rather than a moving mechanical antenna. The antenna's radiated beam is moved electronically.

elevation angle The angle between the horizontal plane and an airborne target under radar observation.

en-route surveillance radar Long-range radars used by the FAA to control air traffic between airports.

exponential position indicator Radar indicator used with precision approach control radar systems. Provides the control with azimuth and glide slope, or descent information.

fast time constant circuit A circuit that is used to compensate for excessively large blocks of video that are capable of masking weaker targets. It is a differentiator that leaves only the leading edge of each target.

feed horn A horn radiator is used to either feed a radar reflector, or intercept the RF energy collected by it. It matches the impedance of the waveguide to that of the antenna system.

frequency agility The ability of a radar system to rapidly change frequency in a predetermined pattern to counter jamming.

frequency modulation The frequency of the carrier voltage is varied by the frequency of the modulating signal.

fast time constant (FTC) A circuit found in the receiver that is used to overcome the effects of heavy clouds associated with weather fronts, jamming, or exceptionally large targets. It is actually a differentiator circuit that passes only the leading edge of the received radar echoes.

frequency spectrum The entire range of frequencies contained within the output pulse of an RF generator.

graceful degradation In a solid state summed high-powered output module, the failure of any one device causes only a minor decrease in output power.

ground clutter Unwanted energy returned from unwanted objects surrounding a radar system. Examples would include vegetation, buildings, and topographical features.

ground controlled approach radar A land-based precision approach radar that is used to provide landing instructions to aircraft during periods of reduced visibility.

gun effect Is when RF oscillations are present in a biased piece of N-type gallium-arsenide semiconductor material when exposed to a 3-kV electric field.

gun oscillator A semiconductor RF oscillator that uses the Gun Effect to operate.

hard tube modulator A radar modulator that uses a vacuum electron tube as a driver to form the modulation pulse.

helix A single-layer coil that is formed into a spiral pattern.

heterodyne The action of combining two input signals and obtaining four output frequencies—the two original frequencies, the sum of the two frequencies, and the difference of the two frequencies.

homodyne The method of reception where the incoming RF is zero beaten with the local oscillator signal. A zero beat condition is when the two frequencies or their harmonics are equal.

homogeneous A liquid, gas, solid, or field that is uniform in composition.

horizontal beam width An antenna's beam pattern measured in the horizontal plane.

hybrid ring A circular waveguide section with four input/output branches. Energy can be transferred from any one input to any two of the three outputs.

identification friend or foe (IFF) Also called secondary radar, it is a secure method to automatically identify an unknown ship or aircraft. A master transmitter interrogates a slave system mounted on a ship or aircraft. The slave system responds with a preselected code, which is received and decoded by the master system. The result is displayed on a radar indicator along with the echo from the suspect vessel.

IF amplifier A narrow bandwidth amplifier section that is tuned to one of the four frequencies produced by the mixer stage.

indicator Electronic equipment that is designed to provide a visual means of displaying target information extracted from the received RF energy.

interrogator The master portion of an IFF system. It is usually co-located with a radar system. It transmits an interrogation pulse train whenever the radar transmitter fires. Any IFF-equipped aircraft within range responds with a coded reply.

keep-alive voltage Voltages applied to TR and ATR tubes to decrease the amount of time needed to ionize the tubes when the transmitter fires.

klystron Multicavity RF generator that is based on velocity modulation to produce a power gain.

line pulsing modulator A modulator circuit that stores energy and develops pulses in the same circuit, usually a PFN.

lobe The beam pattern for a radar antenna. It is the area of the greatest concentration of RF energy.

lossy A conductor having a high rate of attenuation per unit of length.

magic T A four-branch microwave waveguide assembly that is used to couple received RF energy from the antenna system to the receiver. Precision dimensions and matched crystals connected to a balanced transformer couple the RF energy into the receiver.

magnetic field The free space surrounding a magnetized body where a magnetic field can influence objects.

magnetron An RF generator that is based on the interaction of electrons under the combined influence of an electric and magnetic field.

marine radar Surface search radars installed on board small craft and ships for navigation and collision avoidance purposes.

mean time between failures (MTBF) The amount of time that an electronic or mechanical system should function normally between component failures that interrupt operation.

modes The operational phases of a radar system.

modulation The combining of two signals in such a way as to have the voltage of one signal varied by the other. The carrier frequency is lower than the modulating signal.

monostatic A radar system that has a co-located transmitter and receiver.

moving target indicator (MTI) Electronic circuitry that enables a radar system to display moving objects and ignore stationary ones.

noncoherent A waveform in which frequency and phase are not precisely aligned.

parabola A curve in which the locus points are equidistant from a fixed point and a fixed straight line. The fixed point is the focus and the fixed straight line extends from the focus.

parabolic reflector A radar reflector having the shape of a parabola that is capable of concentrating and focusing RF energy.

peak power The maximum value of output power produced by a radar transmitter.

period The duration of one complete cycle of a waveform.

persistence The amount of time that video will glow and remain visible on the face of a CRT.

phanastron A variable sawtooth generator circuit often used in radar indicators to produce sweeps and cursors.

phased array An antenna composed of separate elements that can radiate RF energy in or out of phase with one another. The resulting radiation pattern can be computer controlled to compensate for local conditions.

planned position indicator (PPI) A radar indicator that provides a 360-degree view around the radar system. The radar position is the center of the display. Target position is referenced to its position as compared to the antenna.

precision approach radar A three-dimensional radar that is used to accurately position a landing aircraft in relation to the runway or flight deck.

polarization The direction of the electric lines in radiated RF energy. In horizontal polarization, the electric lines are parallel to the earth's surface. In vertical polarization, the electrical lines are 90 degrees from the surface of the earth.

power gain The ratio of the power radiated by a radar antenna as compared to a known reference.

pulse doppler A doppler radar with a reference pulse to allow for the accurate measurement of range to targets of interest.

pulse modulation A form of modulation in which pulses are the modulating signal.

pulse repetition frequency (PRF) The rate at which a radar transmitter radiates pulses of RF energy and is measured in hertz, or cycles per second.

pulse repetition rate (PRR) The output pulse rate of a radar transmitter.

pulse repetition time (PRT) The length of one complete output cycle of a radar transmitter. Typically it is measured from the leading edge of the transmitted pulse to the leading edge of a second pulse.

pulse width The duration, or horizontal dimension of a pulse.

propagation The extension of energy into and the resulting movement through space.

Q The figure of merit of an inductor, capacitor, LC circuit, or cavity.

radar mile The time it takes RF energy to travel from a radar, strike a target one mile away, and be reflected back to the antenna system. The time interval for two-way travel is 12.36 microseconds. A radar mile is said to be 12.36 microseconds.

radiation loss Energy loss in a conductor caused by radiation to free space.

radio frequency interference (RFI) Unwanted electrical noises that occur in RF amplifiers and detectors.

radome An electrically invisible shell that surrounds a radar antenna to protect it from the environment. Typically constructed from plastic or fiberglass and found in aviation, shipboard, and extremely harsh installations.

range resolution The ability of a radar to distinguish between two close targets on the same bearing, but different ranges.

range ring Concentric rings on a radar indicator to aid the operators in determining the range to a target. The distance between the range rings is dependent upon the maximum indicator range selected.

reactance The opposition to ac current flow from capacitance, inductance, or a combination of the two. Effect is determined by the frequency of the applied ac.

receiver time The time period when the transmitter is inactive and the receiver is processing returned RF energy from targets.

recovery time (RT) The time interval between the end of the transmitted pulse and when the receiver begins to receive returned energy from the antenna system.

reflector A smooth surface, either metal or metal mesh designed to reflect RF radiation.

resolution (radar) The degree to which two adjacent targets can be differentiated. A radar system has both range and bearing resolution characteristics.

resonance The condition in which the natural response frequency of a circuit or cavity corresponds with the frequency of an applied signal.

skin effect The tendency of high-frequency current to flow along the outer surface of a conductor.

spark gap An electronic component consisting of two metal electrodes separated by a small air gap. The application of high voltage causes a spark to jump the gap.

sensitivity time constant A special circuit found in radar receivers to decrease gain immediately after the transmitter fires the RF pulse. Gain is gradually increased as a function of time. The effect is to decrease gain close to the antenna and gradually increase it with distance. It eliminates the problem of close, large targets saturating the receiver circuits, obliterating small targets.

superheterodyne The phenomenon that results when a lower frequency is mixed with a higher one. The result is the two original frequencies, the sum of the two frequencies and the difference between the two frequencies.

surface search radar A surface search radar is optimized to detect targets on the surface of the ocean.

sweep The action of moving an electron beam across the face of a cathode-ray tube. A radar sweep is the electron beam that illuminates the video on a radar indicator.

terminal area radar An airport radar that controls aircraft in the immediate vicinity of the runways. It can also be called an airport surveillance radar and usually has a sixty-nautical-mile range.

three-dimensional radar A radar that is capable of providing the bearing, range, and altitude of a target.

thyratron A gas-filled vacuum tube that functions as a high-voltage, high-current electronic switch. Operation is identical to that of an SCR.

TR A gas-filled tube, or spark gap, that is used as an electronic switch in a duplexer.

trace A moving electron beam across the face of a CRT that illuminates the video returned by targets.

tracking radar A radar that is designed to provide continuous range, bearing, and altitude information to a target of interest. The system typically keeps the target centered in the antenna beam pattern. Similar in concept to a gun sight.

transmission line A single conductor or group of conductors used to route electrical energy from one point to another.

transponder The beacon portion of an IFF system. When signalled by an interrogator, it will respond with a coded reply suitable for display on a radar indicator.

transverse electric In a waveguide, the mode of propagation where the electric lines are perpendicular to the direction of wavefront transmission.

Traveling Wave tube A microwave tube comprised of an electron gun, helical transmission line, collector, input stage, and output stage. An RF signal is coupled into the helix, which parallels the electron beam. Interaction between the two provide for a power gain.

tube replacement Numerous solid-state devices are summed to obtain a high-powered output module for use in a radar transmitter.

two-dimensional radar A radar that is capable of providing only range and bearing information to a target. It can be optimized to track either airborne or surface targets.

waveguide Rectangular or circular metal tubing that is capable of propagating RF energy with minimum losses and reflections. Used to interconnect the RF section of a radar system.

vertical beam width The width of the radiated RF beam in the vertical plane.

video mapper An electronic subsystem that produces a video map to be superimposed on a radar indicator for air traffic controllers. The map will feature runway outlines, air routes, other air fields, topographical features, and other hazards to flight.

VSWR In terms of waveguide, the ratio of the electric field intensity at a maximum point to that of an adjacent minimum point.

white noise Random noise that is uniformly spread across a frequency band.

Bibliography

Dulin, John J., Victor F. Veley, John Gilbert. 1991. *Electronic Communications*. Blue Ridge Summit, PA: TAB Books.

Eaves, Jerry L. and Edward K. Reedy, Editors. 1987. *Principles of Modern Radar*. United States of America: Van Nostrand Reinhold Company, Inc.

Floyd, Thomas L. 1986. *Digital Fundamentals*. Columbus, OH: Charles E. Merrill Publishing Company.

Frieden, David R., Editor. 1985. *Principles of Naval Weapons Systems*. Annapolis, MD: Naval Institute Press.

Freidman, Norman. 1988. *U.S. Naval Weapons*. Annapolis, MD: Naval Institute Press.

Grob, Bernard. 1984. *Basic Electronics*. New York: McGraw-Hill, Inc.

Lynn, Paul A. 1987. *Radar Systems*. New York: Van Nostrand Reinhold Company, Inc.

Miller, David and Lindsay Peacock. 1991. *Carriers*. London: United Kingdom. Salamander Books Ltd.

Nathanson, Fred E. 1991. *Radar Design Principles*. New York: McGraw-Hill, Inc.

Orr, William I. 1989. *Radio Handbook*. Carmel, IN: Howard W. Sams and Company.

Parker, Sybil P., Editor. 1987. *McGraw-Hill Encyclopedia of Electronics and Computers*. New York: McGraw-Hill, Inc.

Polmar, Norman. 1993. *The Naval Institute Guide to the Ships and Aircraft of the U.S. Fleet*. Annapolis, MD: The United States Naval Institute.

Skolnik, Merril I. 1962. *Introduction to Radar Systems*. New York: McGraw-Hill Book Company.

Toomay, J.C. 1989. *Radar Principles for the Non-Specialist*. New York: Van Nostrand Reinhold.

United States Navy. 1988. *Avionics Technician 1 & C*. Washington, DC: Naval Education and Training Program Management Support Activity.

United States Navy. 1986. *Avionics Technician 3 & 2*. Washington, DC: Naval Education and Training Program Management Support Activity.

United States Navy. 1980. *Basic Electronics, Vol. 1*. Washington, DC: Bureau of Naval Personnel.

United States Navy. 1980. *Basic Electronics, Vol. 2.* Washington, DC. Bureau of Naval Personnel.

United States Navy. 1984. *Navy Electricity and Electronics Training Series, Module 17.* Washington, DC: Naval Education and Training Program Development Center.

United States Navy. 1984. *Navy Electricity and Electronics Training Series, Module 18.* Washington, DC: Naval Education and Training Program Development Center.

Index

About the author

Frederick L. Gould (U.S.N. Retired) is currently involved in the space program as an electronics technician with a NASA contractor. He has more than 25 years of experience as a maintenance technician and technical instructor for various military and civilian radar systems. He has also developed technical educational materials for the U.S. armed forces, governmental agencies, and industry.

Other Bestsellers
of Related Interest

Troubleshooting with Your Triggered-Sweep Oscilloscope
—*Robert L. Goodman*
This illustrated guide explains how to set up and use the modern technician's most valuable troubleshooting tool. Electronics repair expert Robert L. Goodman describes in detail the very latest triggered-sweep oscilloscopes and new troubleshooting techniques for rapid, surefire waveform analysis.
0-8306-3891-1 $19.95 Paper